BIOLOGY: BRAIN & BEHAVIOUR

Development and and Flexibility

Springer
Berlin
Heidelberg
New York
Barcelona
Budapest
Hong Kong
London
Milan
Paris
Santa Clara
Singapore
Tokyo

Terry Whatson and Vicky Stirling (Eds)

Development and Flexibility

With 80 Figures

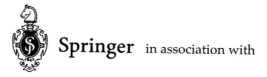 Springer in association with The Open University

Unless otherwise stated, all contributors are (or were at the time this book was written) members of The Open University

Academic Editors

Vicky Stirling
Terry Whatson

Authors

Mike Stewart
Vicky Stirling
Terry Whatson

External Assessors

Richard Andrew, School of Biological Sciences, University of Sussex
  (Series Assessor)
Lynn Bindman, Department of Physiology, University College and Middlesex
  School of Medicine, University College London (Book Assessor)

CIP Data applied for
Die Deutsche Bibliothek - CIP - Einheitsaufnahme
Development and Flexibility/The OpenUniversity. Terry Whatson; Vicky Stirling (eds). - Berlin; Heidelberg; New York; Barcelona; Budapest; Hong Kong; London; Milan; Paris; Santa Clara; Singapore; Tokyo: Springer 1988

Published by Springer-Verlag, written and produced by The Open University

Cover design: *design & production* GmbH, Heidelberg

Printed in Singapore by Kyodo under the supervision of MRM Graphics Ltd, UK.

ISBN 3-540-63794-X Springer-Verlag Berlin Heidelberg New York

This text forms part of the Open University *Biology: Brain & Behaviour* series. The complete list of texts which make up this series can be found above. Details of Open University courses can be obtained from the Course Reservations and Sales Office, PO Box 724, The Open University, Milton Keynes MK7 6ZS, United Kingdom: tel. (00 44) 1908 653231. Alternatively, much useful course information can be obtained from the Open University's website: http://www.open.ac.uk

3.1
SPIN 10654259   #39/3137 – 5 4 3 2 1 0

CONTENTS

PREFACE

Development and Flexibility, like any other textbook, is designed to be read on its own, but it is also the fourth in a series of six books that form part of *SD206 Biology: Brain and Behaviour,* a course for Open University students.

Each subject is introduced in a way that makes it readily accessible to readers without any previous knowledge of that area. Questions within the text, marked with a □, are designed to help readers understand and remember the topic under discussion. (Answers to in-text questions are marked with a ■.) The major learning objectives are listed at the end of each chapter, followed by questions (with answers given at the end of the book) which allow readers to assess how well they have achieved these objectives. Key terms are identified in bold type in the text; these are listed, with their definitions, in a glossary at the end of the book. Key references are given at the end of each chapter, where appropriate. A 'general further reading' list, of textbooks relevant to the whole book, is also included at the end.

The study of the brain and behaviour is an experimental science. This means that it involves the collection of observations, the formulation of specific hypotheses to explain those observations and the carrying out of experiments to test (confirm or falsify) those hypotheses. Throughout this book, these different aspects of the investigative process are emphasized, often through the use of in-text questions in which the reader is invited to engage in the process of deductive reasoning themselves. An understanding of the scientific method, as it applies to the behavioural and brain sciences, is an important aim of this book.

The principal theme of this book is *development.* There are clear similarities between some of the processes of development and those involved in memory formation, which is why both learning and memory also form part of this book. Chapter 1 describes the early growth of the human brain, while Chapters 2 and 3 focus on the growing nervous system and how the appropriate connections are made between neurons, and between neurons and other parts of the body. Chapter 3 also considers the capacity for change in the mature nervous system. In Chapter 4, the output of the developing nervous system—behaviour—is discussed, in particular the role of factors present during development that influence the subsequent growth of the nervous system and hence the patterns of behaviour. The final chapter looks at the biochemical basis of memory.

Before you begin to read this book, there are some important points that you should bear in mind.

1 Experiments on animals

The use of living animals in research is a highly emotive, contentious and political issue. You are no doubt aware of the strong views held by animal liberationists. There is also considerable debate among scientists concerning what kinds of experiments and procedures are acceptable and what are not. Most scientists working with animals seek to minimize any suffering that animals may experience during experiments and each researcher makes his or her own judgement as to whether the suffering caused by an experiment is justified by the scientific value of the results that the experiment yields. The ethics of animal experimentation is not simply a matter of individual judgement, however, but is a matter of concern for society as a whole. In Britain and many other countries, all researchers work within strict guidelines enforced by government; for example, the Home Office licenses all animal experimentation in the UK. Some academic societies, such as the Association for the Study of Animal Behaviour, and many institutions, such as medical schools, have Ethical Committees that oversee animal-based research. In this book, a number of experiments are described; this in itself raises ethical issues because reporting the results of an experiment may be thought to be giving tacit approval to that experiment. This is not necessarily true and it should be pointed out that some of the experiments described were carried out several years ago and a number of them would not be carried out today, such has been the shift in opinion on these issues within the biological community. Paradoxically, certain experiments carried out many years ago, such as those on the effects of maternal deprivation on young monkeys, produced such strong and distressing effects on their subjects—results that were not generally anticipated—that they have had a substantial impact on the kind of experiments that are permitted today.

2   Latin names for species

A particular individual animal belongs to various categories. If you own a pet, it may, for example, be categorized as a bitch, a spaniel, a dog, a mammal, or an animal. Each category is defined by particular features that differentiate it from other, comparable categories. The most important level of categorization in biology is at the level of the species. When a particular species of animal is referred to in this book, its Latin name is also given, e.g. earthworm (*Lumbricus terrestris*).

CHAPTER 1
INTRODUCTION

The previous two books in this series have given you an idea of how the mature nervous system is organized and functions in various animals. This book considers how such a complex system develops. The study of development is important for three reasons. Firstly, development is intrinsically interesting: the problem of how a single fertilized egg cell can become a mature, multicellular animal with its characteristic behaviour has fascinated neuroscientists, developmental biologists, ethologists and psychologists for decades. How does the information contained within the fertilized egg, in combination with internal and external environmental factors, manage to generate an organism of such stunning complexity? Secondly, an understanding of development reveals something about the processes that underlie recovery from injury and regeneration. Current opinion holds that factors operating during development could aid the repair of damaged nerves. The third reason for studying development is that development of the nervous system involves the formation and alteration of synapses. The formation and alteration of synapses also occurs during learning, so an understanding of development may shed light on the processes underlying learning.

Development, like every other topic presented in this course, can be looked at from a number of different levels (Book 2, Chapter 1). These various levels are examined in later chapters, but the book begins with the gross development of the human brain.

1.1   Human brain development

The scale of the problem facing the human zygote is vast. There it is, about the size of a full stop on this page, and yet when the baby is born 270 days later it weighs about 3 kg and is 50 cm in length. The zygote consists of one cell, yet at birth the brain alone contains about 10^{11} (that is, 100 000 000 000) neurons. Not only is there this vast and rapid increase in the number of cells, but there is also the problem of organizing the constantly growing population of cells into organs and tissues that function as a whole.

The very early stages of development are described in Chapter 2. Here the story is picked up at about four weeks after fertilization, when the human brain is starting to develop in the embryo (Figure 1.1 overleaf).

At this stage, when the embryo is about the size of a pea, a groove that runs along the dorsal surface of the embryo closes over and becomes a tube. The tube consists of the cells, called **stem cells,** which will become neurons and glia, and it is from this tube that the brain, spinal cord and other parts of the nervous system are formed. The tube swells at one end—this will become the forebrain—and there are two smaller swellings behind it, which can be identified as the rudiments of the midbrain and hindbrain. At these swellings the hollow core of the tube becomes the ventricles. Within the walls of the tube surrounding the ventricles, stem cells divide to produce neurons and glial cells, which move to form the characteristic structures of the brain; neurons clump together to form distinct structures, and axons begin to link them together. At five or six weeks the tube has to bend to accommodate its own growth within the skull. Two bends occur in very specific places and thereby locate the brain in its characteristic position with respect to the spinal cord.

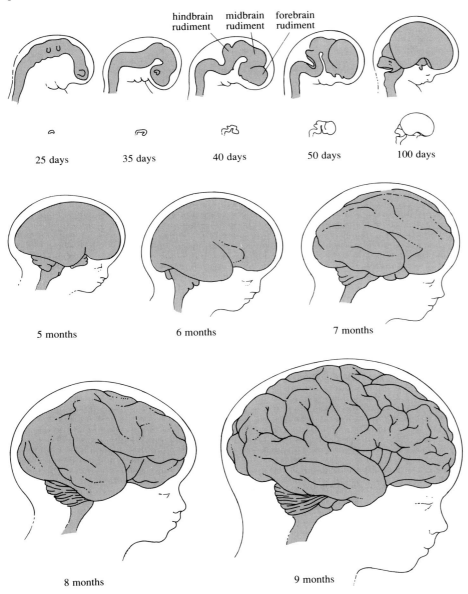

Figure 1.1 Development of the human brain. The diagrams in this sequence are about three-fifths life-size. The first row shows the brains of the second row enlarged for clarity. Note the exceptional growth of the part of the brain which gives rise to the cerebral hemispheres (labelled as the forebrain rudiment at 40 days).

From now on there is a rapid appearance of recognizable structures. Once the embryo has developed a human body plan it is, by convention, called a *fetus*. By the end of the third fetal month, cerebral and cerebellar hemispheres are obvious, and the thalamus, hypothalamus and other nuclei within the brain can all be distinguished. In the following month the cerebral hemispheres swell and extend. By the fifth month the characteristic wrinkles of the cerebral hemispheres begin to appear. Most of the sulci and gyri are apparent by the eighth month of development, although frontal and temporal lobes are still small by comparison with the adult, and the total surface area of the cortex is much below its eventual size.

The full adult complement of neurons in the human brain is present shortly after birth, after which no more are produced. However, glial cells continue to increase in number until adolescence, when the mature adult brain structure is achieved. It is the increase in glial cell number and the growth of axons and dendrites that leads the 350 g newborn brain to become the 1 400 g adult brain.

The rate of increase in the weight of the human brain compared with some other mammals is shown in Figure 1.2 (explained in Box 1.1 below).

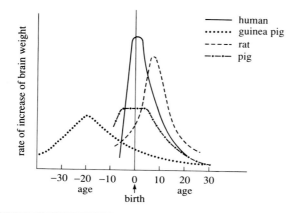

Figure 1.2   The rate of brain growth in four species. Note that the time-scale is different for the different species: months for humans, weeks for pigs (*Sus scrofa*), days for rats and guinea pigs (*Cavia porcellus*).

Box 1.1   Interpretation of Figure 1.2

The growth of the brains of different mammals differ in two major respects: (a) the period of time over which growth occurs—years for humans, weeks for rats, for example; (b) the final size of the brain, 1 400 g for humans, 10 g for rats. The period of time over which the brain grows in different species is accommodated in Figure 1.2 by using a different time-scale for each species—days for rats and guinea pigs (*Cavia porcellus*), weeks for pigs (*Sus scrofa*), and months for humans. Using these different time-scales does not mean that a day for a rat is equivalent to a month for humans. The different time-scales are simply a device for presenting the data on one set of axes.

To interpret the graphs you need to remember that all points on each graph represent increases in the weight of brains: high points are high growth rates and low points are low growth rates. Low points still represent increases in the weights of the brains, but they are small increases.

☐   At what time is the human brain growing fastest—before, at about the time of, or after birth?

■ The period of greatest growth is at or around the time of birth (perinatal) with a considerable amount of slower growth occurring during the first two years.

The period of greatest growth is called the *brain growth spurt*. This observation of the timing of the human brain growth spurt might not be worth making, were it not for the fact that the graphs in Figure 1.2 indicate that the brain growth spurt in different mammalian species occurs at different times. Thus, in the guinea pig it is mostly before birth (pre-natal), whereas in the rat it is mostly after birth (post-natal). The correlation between these observations and the relative states of maturity of these species at birth is most striking—the guinea pig is precocial and the rat is altricial (Book 1, Section 5.1).

A second point that emerges from a comparative study of brain growth spurts is that between species, birth itself is a poor indicator of brain maturity. Where this becomes very important is when the effects on brain growth and development of external agents such as drugs are being investigated and animal models are considered. Applying an external agent two or three days before birth to the rat, pig and guinea pig could give very different answers to the question of whether that agent is safe for human use. A drug that appeared to have no effect on the brain growth of the guinea pig and the rat could still affect brain growth in the pig or the human. Thus, the vulnerability of the brains of different species to the effects of drugs varies at different times with respect to birth. The concept of vulnerability is dealt with in Chapter 4.

This brief description of the gross development of the human brain did not consider the numerous underlying phenomena involved, and said nothing of how neurons develop. The next section moves down a level to look at neuronal development.

1.2 Development of neurons

All the cells in a developing organism are genotypically alike, yet the multicellular animal is made up of cells that have different structures and functions; that is, they are phenotypically different. The structural and functional complexity of the mature animal is not present in the zygote in miniature form, simply ready to enlarge as the animal gets bigger. It emerges through successive small changes as cells access different sub-sets of their genetic information and differentiate.

In Book 1, Chapter 5, Waddington's analogy of development being like a ball rolling down a series of gullies was introduced. Like the rolling ball, development progresses inexorably, and as it does so the environment changes. What the analogy leaves out is the importance of *interaction* in development. There is interaction at all levels—between cells, between groups of cells and, of course, between the organism and its environment. As the animal develops, so the range of stimuli that influence its development increases. To start with, the egg cytoplasm interacts with the egg nucleus; then cells interact with other cells. In the nervous system, neurons make synaptic contact with, and interact with, their targets—that is, the structures that they innervate, such as other neurons, muscles and glands. Finally, the nervous system is shaped by experience.

Neurons grow and mature. They differentiate and move to the right place and then become connected to their appropriate targets. One of the first signs of maturation of a neuron is the outgrowth of its axon, which somehow must reach its target during development. These issues are dealt with in Chapter 2.

Once the axon has reached a target, synapses are formed and communication can begin in earnest. Functional interaction—the ability of one neuron to influence another—plays a very big role in defining the connections between neurons. Functional interaction determines whether synaptic contacts are retained or lost, and even whether the neuron itself will survive or die. It has been realized for a long time that the amount of genetic information contained in the DNA of the fertilized egg is not enough to specify all the connections between all the neurons in the nervous system. Instead, the exquisite accuracy of neural connections is defined through neural activity. Chapter 3 presents evidence related to these issues.

1.3   Specificity versus plasticity

In Section 1.1 it was stated that the various sulci and gyri that can be observed in the adult brain are present by the eighth month of development. If these sulci and gyri are present from before birth to adulthood, then does this mean that the structure of the brain does not change throughout life? Put another way, are the architecture of the brain and the neuronal connections in it laid down or specified early in development? This notion of **neuronal specificity** can be stated more precisely as, *particular neurons are always connected to particular targets*; each neuron has a specific target that it innervates. A further question arises from the concept of neuronal specificity. If the structure of the nervous system is specified, is it specified by the genetic material?

The answer to both these questions about specificity is a resounding 'sort of'! The structure of the brain is specified (after all, if the structure did change, the sulci and gyri, for instance, would be found in different places in different people, and in different places in the same person at different times, but this is not the case). Such specification of structure is even more apparent in the locust, described in Book 2, Chapter 7, where the position of each individual neuron varies little from animal to animal. However, the structure is specified only at the gross level. As you will see in Chapters 3 and 4, the precise way in which individual neurons connect and communicate depends critically on circumstance. Such **neuronal plasticity** means that the connection between a neuron and its target is neither specified nor fixed. Indeed, as will be discussed in Chapter 5, it is plasticity at the synapse that is the basis for learning, and learning is possible throughout life. Another aspect of plasticity is the ability of some mature axons to grow and form functional connections after they have been damaged, an ability discussed in Chapter 3.

The answer to the second question about whether the architecture is specified by the genetic material, whether somehow the information contained in the genetic material dictates the structure of the brain, is also only partly correct. As noted above, there is not enough genetic information in any organism's DNA to specify the connections between all the neurons in its nervous system. However, the genetic material has a considerable influence over the eventual gross structure of the nervous system, with, as ever, the environment exerting its influence too. Thus, the environment in which axons grow, their targets and the activity of neurons all influence the eventual connections they make (Chapter 3), just as the environment influences the way the nervous system functions (Chapter 4).

Clearly, in the early stages of development, the organism is undergoing more change and is more plastic, than at later stages of development. As the nervous system matures, so its capacity for plasticity becomes less and less. (Perhaps this is another difference between altricial and precocial species; altricial species have a more plastic nervous system when born, and are able to adapt to circumstances during development in a way that precocial species cannot; see Book 1, Section 5.1.) Understanding these changes in plasticity and their causes is a major challenge for developmental neurobiology, and is a major theme of this book.

Summary of Chapter 1

On the surface of the zygote a tube forms. The stem cells of which the tube is composed divide, migrate and differentiate. As they differentiate, they interact with each other and access different parts of their genetic material to become the neurons and glia of the brain and nervous system. The neurons grow axons which make contact with particular targets. Such contacts are permanent but not immutable. Given the right circumstances, contacts can change, for the nervous system retains an inherent plasticity.

Objectives for Chapter 1

After reading this chapter, you should be able to:

1.1 Define and use, or recognize definitions and applications of, each of the terms printed in **bold** in the text.

1.2 Give a very brief account of the development of the human brain.

Questions for Chapter 1

Question 1 (*Objective 1.2*)
Where does cell division occur in the brain?

Question 2 (*Objective 1.2*)
Look at the line for human brain growth in Figure 1.2. (a) At what age does the line stop? (b) At the age where the line stops, has the brain stopped growing?

Question 3 (*Objective 1.1*)
What is the difference between neuronal plasticity and neuronal specificity?

CHAPTER 2
BIRTH AND GROWTH OF NEURONS

2.1 Introduction

This chapter and the next deal with how the nervous system grows: how cells come to be neurons, how neurons move through the embryo and stop at some appropriate position within the embryo, how axons establish appropriate connections with other neurons and with muscles and glands. Chapter 2 covers the process up to and including axon growth, and Chapter 3 deals with synapse formation.

There are six main phases in the development of the nervous system. Although in general these phases are consecutive, they often overlap in time. For example, many neurons grow (that is, produce an axon) while they are migrating. Also, many of the phases may be occurring simultaneously in different parts of the nervous system. The six phases are:

1   A *proliferation phase*, during which cell division occurs, resulting in an increase in cell number.

2   A *phase of differentiation*, during which cells form the characteristic features of neurons; some of these neurons continue to proliferate.

3   A *migration phase*, during which neurons move from their place of origin to their final position.

4   A *growth phase*, during which neurons grow axons.

5   A phase during which the axons establish *synaptic connections* with targets (for example sense organs, muscles or other neurons), and themselves become the targets of other neurons.

6   A *modification phase*, during which the connections formed during the growth phase are modified, with some being removed (involving the death of up to 50% of the neurons originally generated) and some being strengthened. This modification, or fine tuning, depends critically on the establishment of functional connections.

What are the factors that control these phases and, more particularly, how are the processes that occur in each phase controlled? Why does a neuron differentiate to become a specialized cell, for example a Purkinje cell rather than a granule cell in the cerebellum (Book 2, Section 8.8.8), a rod rather than a cone? What guides a growing axon so that it connects to one particular target rather than another? Do sensory neurons find an appropriate target or are neurons non-specific until they meet a target, becoming sensory if they meet, for example, a Paccinian corpuscle and motor if they meet a muscle? These are the kinds of question to which developmental neurobiologists seek answers, and it is these questions that are addressed in Chapters 2 and 3.

Before these questions of detail are considered, the next section deals briefly with the overall early development of an organism, as it changes from a zygote to an embryo, with special emphasis on the nervous system.

2.2 Early embryonic development

Figure 2.1 shows drawings of various vertebrate embryos at three ages. Notice how similar the embryos look; they only become recognizably different close to the end of development. This observation is very important because it suggests that the early stages of development of all vertebrates are similar, so that the processes involved are probably common to all vertebrates.

Next to some of the drawings are numbers preceded by an E. These numbers indicate the *embryonic age*, that is, the number of days that have elapsed since

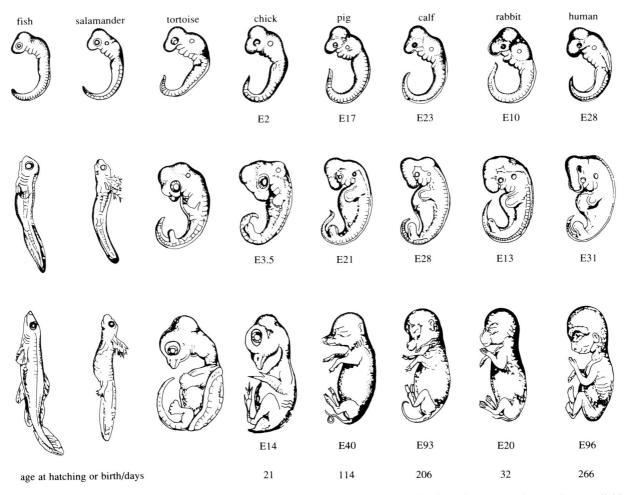

	fish	salamander	tortoise	chick	pig	calf	rabbit	human
				E2	E17	E23	E10	E28
				E3.5	E21	E28	E13	E31
				E14	E40	E93	E20	E96
age at hatching or birth/days				21	114	206	32	266

Figure 2.1 The appearance of vertebrate embryos at three stages of development. Embryonic ages are shown where available.

conception (= fertilization). For instance, E20 means embryonic day 20. Time *after* birth or hatching is given the prefix P, for post-natal. This code is a convenient way of noting the developmental age of an animal.

☐   Compare E28 in the human with E20 in the rabbit. What does this tell you about the relative rates of development?

■   The E20 rabbit has recognizable limbs and body, whereas the E28 human does not. So the rabbit must develop much more quickly than the human.

Often the word *fetus* is applied to older mammalian embryos (there is no clear distinction between *embryo* and *fetus*). In this text the word *fetus* will only be applied to human embryos after they have developed the main body plan and limbs (that is, by E96 in Figure 2.1).

Knowing the embryonic age of an animal does not necessarily give any indication of the state of maturation of the embryo because of the enormous variation in the rate of development. These differences in the rate of development also mean that there are differences in level of maturity at birth.

Figure 2.2 is a series of drawings showing the main stages of development in the frog *Xenopus laevis* (also, confusingly, known as the South African clawed toad).

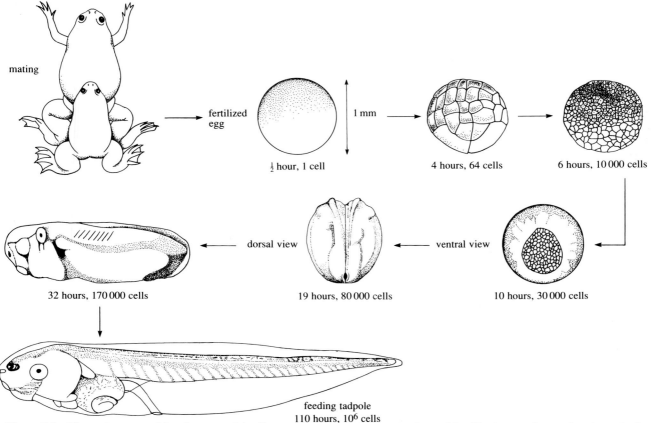

Figure 2.2   The main stages of development of the *Xenopus laevis* tadpole from mating and fertilization to a free-swimming tadpole—about 4½ days.

Frog embryos (frog spawn) are easy to obtain and observe, and their simplicity makes description of the basic events taking place much more straightforward than in mammals. The very first division of the fertilized egg is particularly important; it defines the midline of the embryo. As both sides of the embryo are the same, it is bilaterally symmetrical. The position of this first division is dictated by the entry point of the sperm into the egg and the distribution of the heavy yolk in the egg.

The resulting two cells each divide again, making four cells. These cells continue to divide, eventually making a hollow ball of cells. At about 10 hours (Figure 2.2) the cells then go through a complex series of movements in which the ball effectively turns in on itself through a region called the *blastopore* to produce a structure consisting of three cell layers. Each of these cell layers develops into a particular class of cell. The outermost layer of cells, the *ectoderm*, forms the skin and nervous system, the middle layer, or *mesoderm*, forms the muscles of the body and skeletal structures, and the innermost layer, or *endoderm*, forms the internal organs and the gut. Already at this stage there is a clear difference in the way different classes of cells grow and mature; in other words, differentiation has already started.

A sequence of cell divisions in the ectoderm results in the formation of a hollow tube (the *neural tube*), which will develop into the central nervous system as shown in Figures 2.3a and b. Notice that at a slightly later stage there are three distinct bulges at one end of the neural tube (Figure 2.3c). These bulges are the future forebrain, midbrain and hindbrain (Section 1.1).

The forebrain bulges outwards on each side, and, where it meets the overlying ectoderm, two cup-shaped rudimentary eyes form. Similar interactions between the neural tube and the ectoderm produce structures that will become the nose and ears.

☐ What do the eyes, nose and ears have in common?

■ One thing they have in common is that they are all sense organs.

Note that the nervous system, and many peripheral sense organs, are all derived from the outermost layer of cells, the ectoderm. This point is discussed further in Section 2.3.

Meanwhile, further caudally, the spinal cord enlarges, and mesodermal cells migrate to form regularly spaced blocks of tissue down each side. These blocks of muscle tissue are called *somites*; they will later form the cartilage and bone of the backbone, and also the muscles of the body. At 110 hours after fertilization, the frog embryo is already recognizably a vertebrate. It has a hollow dorsal segmented nerve cord and a three-lobed brain with specialized sensory organs, which now start to differentiate.

The early development of all vertebrates follows this general plan, but the details of the shapes of the cell masses may differ. Thus, in birds the embryo develops as a flat sheet of cells floating on the yolk of the large egg, so that the exact process forming the three cell layers is rather different, but the end result is the same, the formation of a hollow dorsal neural tube. The beginnings of the future limbs are visible as small mounds of tissue called the limb buds (Figure 2.4).

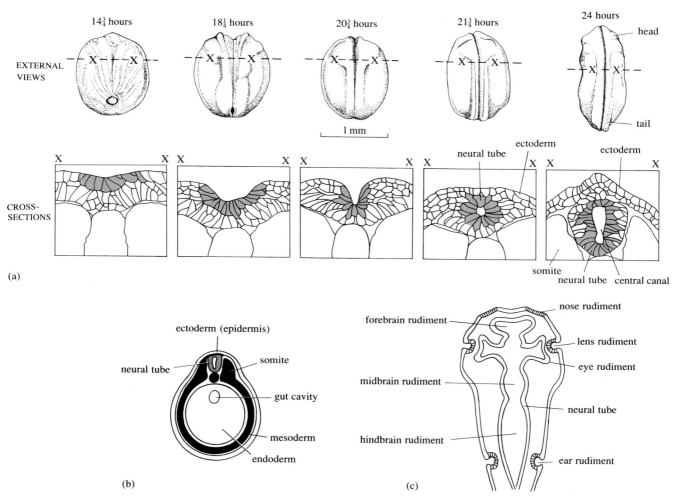

Figure 2.3   (a) Neural tube formation in *Xenopus laevis*: the shaded flat sheet of cells in the ectodermis at $14\frac{3}{4}$ hours later becomes a groove, and by 24 hours has developed into a tube. The upper row shows a dorsal view of the embryo, whereas the lower row shows transverse sections made from X to X at the position indicated by the broken line. (b) Diagram of a transverse section through the whole embryo after the formation of the neural tube at about 24 hours after fertilization, showing the three cell layers. (c) The development of the regions of the brain and the sense organs at about 24 hours after fertilization, shown in a horizontal section.

During this early phase of development, the basic plan of the vertebrate body is formed from the division and migration of cells. As development proceeds, the cells differentiate to become different from one another. This glimpse of the early stages of life of the embryo has shown how the three phases—cell proliferation, migration and differentiation—are involved in development. Why some cells continue dividing throughout life (for example liver cells, or cells in the immune system), whereas others cease being able to divide early in life (for example neurons) will not be considered further here, though it is worth noting that each region of the central nervous system has its own particular pattern and timing of cell proliferation and migration. In mammals some areas of the brain, such as the cerebellum and hippocampus, continue to generate neurons after other areas have reached their final complement of neurons. Cell migration shares many features with axon growth and will not be considered separately; cell migration will be discussed with axon growth in Section 2.3.4.

The next section deals with some factors that control cell differentiation in the embryo.

2.3 Cell differentiation

In the previous section it was stated that the nervous system is formed from ectodermal cells and not from mesodermal or endodermal cells. A reasonable question to ask is: why not? Why can't mesodermal cells become neurons or glial cells? And conversely, why can't ectodermal cells become muscle cells? The answer has to do with differentiation. Essentially, ectodermal cells are already differentiated to the extent that not all developmental paths are open to them; some paths, for example those leading to muscle cells, are closed.

The muscle cells and neurons (indeed all the cells of an animal) contain the same DNA. The cell types differ in which parts of the DNA are translated into protein (Book 1, Section 3.2). The proteins that a cell contains determine its structure and function. Differentiation is the process by which some parts of the DNA become available for translation into protein, whereas other parts of the DNA become unavailable. In ectodermal cells, that part of the DNA required for a cell to become a muscle cell is unavailable, so ectodermal cells cannot become muscle cells; that developmental path is closed.

The purpose of this section is not to unravel how the DNA is made available or unavailable for translation, but to explain how cells come to have different factors affecting their DNA. After all, all the cells of an embryo are descended from the same zygote, and they all have the same DNA. Somehow, therefore, different cells must have different factors affecting their DNA so that different parts of the DNA are translated. How does this happen?

2.3.1 Cell lineages

Imagine trying to divide some minestrone soup into two equal portions in one go. You could probably get the volumes about right, and maybe even the number of bits, but there are bound to be more noodles or carrots in one portion than the other. A similar problem faces the zygote. When it divides, there will be equal numbers of chromosomes and roughly equal numbers of the major structures (for example mitochondria) in each daughter cell, but particular substances will not be evenly distributed: one daughter cell will have more of one substance than the other daughter cell. Therefore the two daughter cells are not identical, they are just very, very similar. When the daughter cells divide, their own daughters will not be identical either. Once the cells of the embryo have divided a couple of hundred times these tiny differences will be magnified and the cells will contain different factors. Could this notion of unequal division explain what is known about cell differentiation?

It is possible to study which cell divided to produce which other cells by injecting a cell with a marker, such as a non-toxic dye or a radioactive substance (label). As

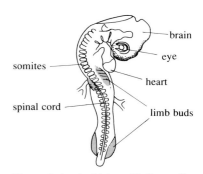

Figure 2.4 A chicken (*Gallus gallus domesticus*) embryo at E2½ days.

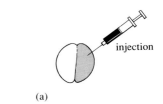

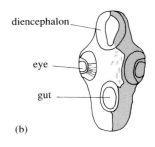

Figure 2.5 (a) Injection of horse radish peroxidase into one cell of a *Xenopus* embryo at the two-cell stage. (b) Drawing of a transverse section at a later stage, showing the position of HRP-containing cells. The area of cells containing HRP is shaded grey.

the injected cell divides, the marker is distributed between the two daughter cells. In this way, all the progeny of the injected cell contain the marker and so the **lineage** of the injected cell can be traced.

The progeny of one cell is shown in Figure 2.5. The marker (horseradish peroxidase, HRP) was injected into a single cell in the embryo at the two-cell stage, as in Figure 2.5a. The injected cell continued to divide and all its progeny can be seen at later stages because they contain HRP (Figure 2.5b). Lineage analysis can be used to discover the origin of particular cell types. Such analyses reveal that the origins of particular cells are remarkably similar from one animal to another of the same species.

Notice that most of the filled cells (that is, those containing HRP) are confined to one half of the embryo.

☐ What does this observation suggest?

■ As most of the filled cells are confined to one half of the embryo, it would suggest that each cell at the two-cell stage generates half the embryo.

The evidence for cell lineage controlling differentiation is even clearer in the nematode worm *Caenorhabditis elegans*. The small size (only 0.5 mm long with a total of 959 cells) of this creature makes it possible to follow cell lineages in living embryos (Figure 2.6). Notice that some cells are derived from only a few cell divisions, whereas others are derived from many. In the nervous system of this animal the fate of many neurons seems to be controlled by cell lineage alone. This lineage is invariant: a particular cell in a particular species of nematode worm will always divide to form the same types of differentiated cell occupying the same position in every member of that species. Removal of a particular cell in the early embryo results in the predictable loss of particular tissues or structures; for example, removal of cell A in Figure 2.6 results in the loss of structure B.

So it is possible for cell lineage to determine how cells differentiate. However, this is far from the whole story, as the next four sections reveal.

2.3.2 Regulation

In many organisms, if the two cells from the first cell division are separated using a loop of fine hair, each will go on to form a complete embryo. This separation sometimes happens naturally with human zygotes, to give rise to identical, or monozygotic, twins. Although each cell would normally make only half an embryo, when they are separated, they each repeat that first division and generate a complete embryo. Any cell isolated from an embryo up to the eight-cell stage has this remarkable ability of **regulation** for reconstituting a complete embryo. A single cell isolated from the embryo after the eight-cell stage will not generate a whole animal; part of it will be missing.

☐ How do these observations accord with the cell lineage theory?

■ Up to the eight-cell stage (that is, the third cell division; see Figure 2.7), the individual cells are still sufficiently similar so that any one of them can develop into a complete embryo. However, if cell lineage were the only

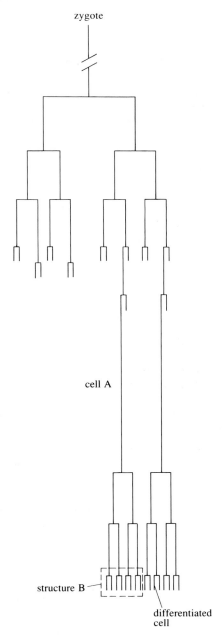

Figure 2.6 Part of the cell lineage of the nematode worm *Caenorhabditis elegans*. Each vertical line represents a cell; each horizontal line represents a cell division.

mechanism of differentiation, then a cell that had been removed from the eight-cell embryo would not replace the seven missing cells before continuing to divide and develop; it would simply divide and differentiate from the stage it had reached, but would not develop into a whole embryo.

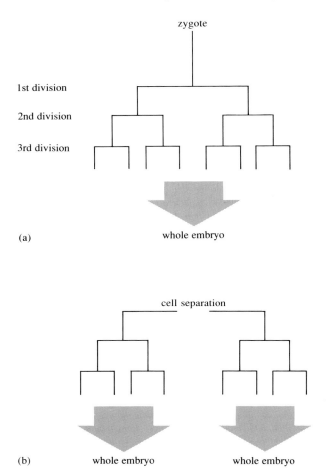

(a)

(b)

Figure 2.7 (a) Normal cell division from the zygote leads to a complete embryo. (b) Cells separated after the first division and before the fourth division will develop into two whole normal embryos—an illustration of regulation.

This shows that the early embryo is not made up of a defined set of cells, each of which follows its own predetermined course of differentiation. Throughout development the interactions between cells play a very important role in controlling differentiation. The inability of cells isolated later than the eight-cell stage to form complete embryos also illustrates another fundamental developmental rule, namely that, as development proceeds, the capacity of cells to change their course of development decreases.

Regulation is not confined to the pre-eight-cell embryo. If part of the forebrain is removed at, say, E13 in the rabbit, then the adult animal will have a normal brain: the remaining cells adjust their development to replace the missing cells. However, if the whole forebrain rudiment is removed, or it is removed at too late a stage, the animal will be missing its forebrain and diencephalic structures. This kind of experiment suggests that particular populations of cells act together to form specific regions of the brain, and that damage to part of such a cooperative system will result in regulation and regeneration of the damaged region. However, if the whole ensemble is removed, it cannot be replaced by other cells in the vicinity.

The phenomenon of regulation means that cells must be able to detect and respond to the presence of adjacent cells.

2.3.3 Intercellular messages

Consider the following transplantation experiment performed in 1924 by Hans Spemann and Hilde Mangold in Germany. The transplantation was done on the axolotl *Ambystoma mexicanum* (a kind of salamander) at a very young age, that is, prior to neural tube formation. They removed a small piece of tissue from the blastopore region of one embryo, the donor, and transplanted it into ectodermal tissue of the ventral belly region of a second embryo, the host, of the same age. Some time later, a second neural tube developed on the host, forming a sort of Siamese twin embryo, as shown in Figure 2.8. They used an albino (white) axolotl as the donor and a pigmented axolotl as the host. The second embryo turned out to be remarkably complete and was mainly composed of pigmented (host) cells.

☐ What had happened to the host cells under the influence of the donor tissue?

■ The host cells that would have differentiated into belly structures changed their developmental pathways and formed the tissues of the second embryo; the donor tissue had 'persuaded' or *induced* the host belly cells to form a secondary dorsal neural tube.

This demonstration of **induction** suggests that the differentiation of cells involves some kind of interaction between cells. This interaction involves 'messages', presumably chemical messages, passing between cells. Inspired by these experiments (for which Spemann received the Nobel Prize for Physiology or Medicine), many embryologists have carried on trying to identify the substances that could be responsible for induction, but so far with little success.

Further transplantation experiments revealed the following:

1 Donor tissue from regions other than the blastopore does not induce host belly cells to form a second embryo.

2 Donor blastopore tissue transplanted next to host mesodermal or endodermal tissue (as opposed to ectodermal tissue) does not induce a second embryo.

3 Transplantation of tissue at a later stage of development, but from the same region of the blastopore, to host ectoderm does not induce a second embryo.

☐ What does result 1 suggest?

■ That the chemical messages produced by the blastopore region are different from those produced by other regions. Only the messages produced by the blastopore region can induce host belly cells to form a second embryo.

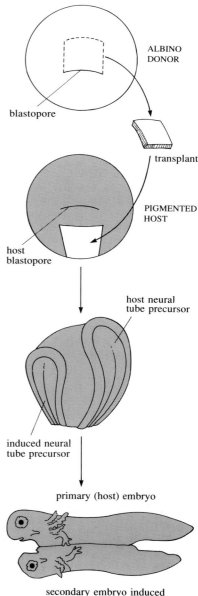

Figure 2.8 Transplantation at about E6 hours of a particular small part of the blastopore region of an albino axolotl into the ventral region of a pigmented host embryo results in the induction of a second neural tube and the formation of a secondary embryo.

☐ What does result 2 suggest?

■ Mesodermal and endodermal tissue do not respond to the messages produced by the blastopore in the same way as ectodermal tissue. Indeed endodermal tissue does not respond at all.

Thus, induction is not a one-sided process; there must be a tissue sending a particular message, and a tissue able to receive and respond to that message.

☐ What does result 3 suggest?

■ Result 3 suggests that the blastopore region only produces the message for a limited time.

By using transplants and hosts of different ages, it has been established that there are critical times at which the blastopore region is capable of inducing, and critical times at which the ectodermal belly tissue is capable of responding.

These transplantation experiments show that during normal development the future organization of the nervous system is established partly through the inductive influence of the blastopore cells on the overlying ectodermal cells. Further interactions between cells in different parts of the nervous system are responsible for laying down regional or localized patterns. Transplantation experiments have also shown that, as cells differentiate, so they become more and more specialized, and are less able to adopt a new course of development when transplanted.

Induction is a general process of development, not confined to the nervous system. The differentiation of all cells depends on signals passing from one group of cells to another, because these signals affect which parts of the DNA are translated into proteins, and hence which proteins are produced in a cell.

The next section considers the process of cell generation in the nervous system.

2.3.4 Cell division in the nervous system

Cells can differ not just in their lineage and the intercellular messages they receive from other cells, but also in their 'birth date'. The birth date of a neuron is the date at which the neuron is formed and after which it ceases to make further cell divisions. Dividing cells in the cortex can be marked using a radiolabel (see Box 2.1). Neurons born at the time of the radiolabel injection contain the most label.

If an embryo is exposed to radiolabelled thymidine (by injection into the mother's bloodstream, for example), the cells that are dividing at the time of exposure will be radiolabelled. Such cells are usually just said to be 'labelled'. Labelling experiments support the earlier histological observations that, in most areas of the central nervous system, the finally dividing cells are found very close to the central canal of the spinal cord or the ventricles of the brain.

Embryonic mice were exposed to radioactive thymidine via injection into their mother's bloodstream at different stages of development. Some were exposed at developmental age E11, others at E13, others at E15 and still others at E17 (mice are born at E21). The newborn mice were killed at P10 and a section of their cortex

Box 2.1 Autoradiography of dividing cells

Thymidine autoradiography provides a direct picture of dividing cells. Thymidine is one of the four DNA bases (Book 1, Section 3.2.2). When a cell divides, it must first make a copy of its DNA. If an animal is injected with thymidine labelled with radioactive hydrogen (radiolabelled thymidine), dividing cells incorporate it into new copies of their DNA. The daughter cells will therefore contain radiolabelled DNA, which can be revealed under the microscope by the technique of autoradiography, the basis of which was explained in Book 3, Box 4.1. Cells that were not dividing when the radiolabelled thymidine was injected do not incorporate it into their DNA. Using radiography, it is thus possible to provide a picture of where cells were dividing.

examined by autoradiography. A summary of the results is shown in Figure 2.9 The radiolabelled thymidine is only present for a short period (it is rapidly lost from the mother's bloodstream), so the cells that are most strongly labelled in the post-natal animal are those that divided for the *last* time when exposed to the labelled thymidine, that is, the new-born neurons. These neurons will contain the label permanently. Neurons that are subsequently derived from cells that took up the label but which continued to divide will contain less label, because the amount of label is reduced at each cell division. These neurons will be weakly labelled at P10. Thus, neurons born at the time of injection can be distinguished from later-born neurons.

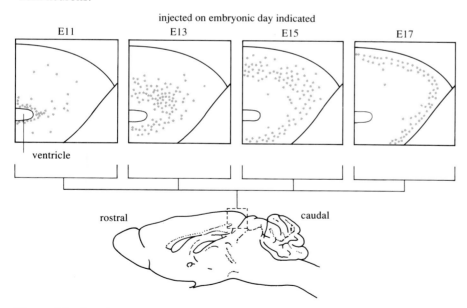

Figure 2.9 A series of drawings of the location of labelled neurons in the cerebral cortex of four P10 mice that had been exposed to radiolabelled thymidine at E11, E13, E15 and E17, respectively, by injection into the mother's bloodstream. Only those cells that divided for the last time when the thymidine was injected are shown. The lower diagram indicates the part of the brain from which the section was taken.

☐ Given that the new neurons are born along the wall of the ventricle, what does the distribution of neurons in the cortex of the P10 animals shown in Figure 2.9 reveal about what happens to cells following their final division?

■ The neurons born at the earliest times are all found in the deeper layers of the cortex, near the ventricle, whereas the later-born neurons occupy the upper layers, furthest from the ventricle. This means that neurons born later must migrate past the earlier-born neurons to their final position.

Notice that all the neurons occupying a particular layer of the cerebral cortex are born at a similar time; they all have the same birth date.

The relative roles of birth date and intercellular messages on cortical cell differentiation have been examined by looking at the axonal projections of neurons in the cortex of normal and 'reeler' mice.

The reeler mouse has a single gene mutation (Book 1, Section 3.2.5). It was identified because it has difficulty in maintaining its balance. This was found to be associated with an abnormal cerebellum. When the development of the nervous system in reeler mice was examined, the pattern was found to be somewhat different from that of normal mice. Using thymidine labelling methods as described above, it was immediately obvious that the pattern of neuronal cell layers by birth date was abnormal in both the cerebellar and cerebral cortices. In the cerebral cortex the earliest-born neurons were found furthest from the ventricle, whereas the later-born neurons were found nearest the ventricle. Somehow the defective allele had affected the migration of the neurons following their final cell division. Histologically it was obvious that the cerebral cortical structure in the reeler mouse was upside down, with the large pyramidal neurons furthest from the ventricle (Figure 2.10).

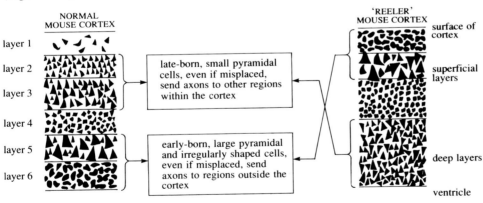

Figure 2.10 Diagram showing the anatomy of the cerebral cortex in a normal and a reeler mouse. The projections of certain neurons are described in the central boxes.

☐ What does this observation reveal about the relative importance of birth date and position following cortical cell differentiation?

■ The differentiation of the cortical neurons depends on their birth date rather than their location. If location were important, neurons nearest the ventricle would have become large pyramidal and irregularly shaped neurons. They actually became small pyramidal neurons.

What about the axonal projections from these neurons? If the large pyramidal neurons are in the 'wrong' place, do their axons also project to the wrong place?

☐   What is the axonal projection of the cortical pyramidal neurons in normal animals: (a) from pyramidal neurons in the deep layers (that is, layers 5 and 6) and (b) from pyramidal neurons occupying more superficial positions?

■   The pyramidal neurons in the deep layers send their axons to other areas of the brain, whereas the smaller pyramidal neurons in the superficial layers connect different regions of the cortex together (Book 2, Section 8.8.1).

When tracers (Book 2, Box 8.1) were injected into the upper layers of the reeler cerebral cortex, axons that projected to brain areas other than the cortex were filled, whereas injection into deeper neurons filled axons that remained within the cortex. These results are shown in Figure 2.10.

Thus, the overall projections were appropriate for the type of neuron, not the position of a neuron with respect to the ventricle and the surface of the cortex.

However, although the projections were unaffected by the depth of the neuron in the cortex, it does appear that the destination of the axon is affected by whether the neuron is located in a motor or a sensory part of the cortex. This was elegantly demonstrated by Dennis O'Leary in St Louis, USA. Using embryonic rats, he transplanted pieces of primary sensory cortex containing large pyramidal cells of layer 5 to the primary motor cortex region and vice versa.

☐   Which cortical cells will be filled (that is, located) by application of tracer to the pyramidal tract?

■   Application of tracer to the pyramidal tract will fill the large pyramidal-shaped cells in layer 5 of the motor, but not the sensory, cortex (Book 2, Section 8.8.1).

The pieces of cortex were marked with a long-lasting dye before they were transplanted so that they could be identified later. When the pyramidal tract was injected with another dye later in development, it filled pyramidal cells in layer 5 of the motor cortex. No dye was found in the cells that had been taken from the motor cortex and transplanted to the region of the primary sensory cortex. Conversely, cells from the primary sensory cortex that had been transplanted into the primary motor cortex were filled. For both types of transplants the cells projected to those areas characteristic of their *new position* in a different region of the cortex rather than to areas characteristic of their *origin*.

This brief account of cell division and migration in the central nervous system has shown that what cells develop into depends partly on their lineage and birth date, and partly on their position.

☐   How could the position of a cell affect its development?

■   Cells in different positions will be in different environments. The surrounding cells will be producing different intercellular messages from cells in other positions.

The immensely complex structure of the brain and spinal cord arises through the interaction of cells, different regions following different programmes of cell division, and migration.

This section has covered phases 1, 2 and 3 of the general scheme of nervous system development presented in Section 2.1, namely the phases of cell division, migration and differentiation, for the *central* nervous system. The next section considers these processes in the *peripheral* nervous system.

2.3.5 Cell division and migration in the peripheral nervous system

The neurons and glia of the peripheral nervous system do not originate from the neural tube, but arise from special cells that collect on either side of the dorsal part of the neural tube called **neural crest** cells (Figure 2.11). These neural crest cells migrate away from the neural tube. Their migration pattern is closely related to the pattern of the somites (see Figure 2.4 for the location of the somites): they only migrate through the rostral half of each somite.

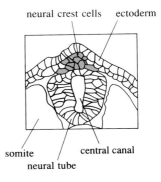

Figure 2.11 The location of neural crest cells with respect to the newly formed neural tube.

The migration patterns and differentiation of neural crest cells have been analysed very elegantly by transplanting neural crest cells from a quail (*Colinus virginianus*) embryo into a chicken host. Provided the transplantation is done early in development, the transplanted cells continue to develop in the new host. The fate of the transplanted cells can be followed because quail cells possess a very distinctive nuclear structure, which is easy to identify histologically. Transplantation experiments in which quail cells are transplanted to different locations in the chick embryo show that the way the cells migrate and differentiate depends on their current location, and not where they were taken from in the donor embryo. This finding is similar to that reported above when pieces of cortex were transplanted in embryonic rats.

Figure 2.12 shows the migration route taken by quail neural crest cells after transplantation into the thoracic region of a chick host of the same age. Neural crest cells migrate to form the segmentally arranged dorsal root ganglia (in the rostral half of each somite); others migrate further to form the ganglia of the autonomic nervous system, and the adrenal medulla and the glia of the peripheral nervous system, or pigment cells in the skin (not shown) and internal organs.

In the peripheral nervous system, the differentiation of cells depends critically on their position, that is, their local environment. As yet, it is not known if cell lineage is involved.

2.3.6 Summary of early development

The nervous system starts off as a simple hollow nerve cord made up of ectodermal cells. After a neuron has been 'born' and has migrated to its final position in the nervous system, how it develops depends both on its ancestry and also on its local environment, that is, on intercellular messages. Both the ancestry and the local environment are able to affect which parts of a cell's DNA are translated into proteins, and hence, how the cell differentiates. The influence of cell-to-cell interaction on development means that, even if the complete genotype were known, it would not allow the exact development of an animal to be predicted.

Such interaction also means that the link between a simple gene mutation (affecting a single protein for instance) and the resulting phenotype is very complex, since what is involved is the effect of the altered or missing protein on a whole hierarchy of subsequent cellular interactions. The projection of the axon, both in the central and in the peripheral nervous system, is affected by the location of the cell body. It is the growth of the axon that is the subject of the next section.

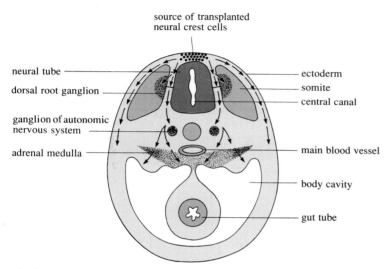

Figure 2.12   Diagram of a transverse section through the thoracic region of a chick embryo, showing the pathways of migration of transplanted quail neural crest cells (arrows). Note that the section is through the rostral part of a somite. The caudal part of the somite, which cannot be seen in the diagram, would be sticking out of the page towards you; the rest of the rostral part would be beneath the page away from you.

2.4   Axon growth

One of the characteristics of the structure of the nervous system is the complex arrangement of connections made by axons and dendrites. There is remarkable similarity in the patterns of axons connecting particular parts of the central and peripheral nervous system among animals of the same species. Often the distance between the cell body of a neuron (for example a motor neuron in the spinal cord) and the point where its axon makes synapses (for example a calf muscle) is considerable. In the embryo, at the time when the axons make contact with their targets (that is, the neurons, muscles or glands with which they make synaptic contact), these distances are much smaller, though some axons still have to grow over hundreds of micrometres or even millimetres. This section describes some of the processes that help to guide axons to their targets during development.

2.4.1   How axons grow

One of the first signs that a cell will become a neuron is the extension of a process, the future axon, from one point on the cell body. This usually happens when the cell reaches its final location in the brain, but can also occur as cells are migrating.

Ramon y Cajal was one of the first neuroanatomists to identify the often flattened structure at the tip of the growing axon (called the **growth cone**) as playing a vital part in the process of guidance towards targets. He also realized that guidance involves dynamic interaction between the growth cone and the environment through which it is growing.

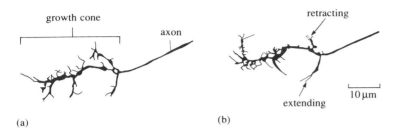

Figure 2.13 Drawings of the moving growth cone of a retinal ganglion cell axon. The cells of a living fish embryo were injected with dye and their growth recorded on video. Drawing (b) was taken from a video image 15 minutes after the one on which (a) was based.

Observations of growing axons (Figure 2.13) show that growth cones are very active structures, continuously making and retracting branches as they extend towards their targets. Seen with a normal light microscope, the growth cone is rather an empty-looking structure containing a few organelles such as mitochondria. Using video techniques, the internal activity of the living growth cone appears as a mass of jostling vesicles. Figure 2.14 shows two video pictures of a growth cone of an *Aplysia* neuron in culture taken 7 minutes apart.

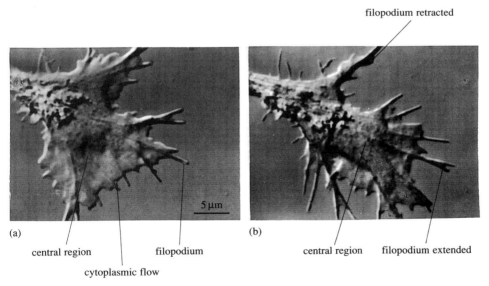

Figure 2.14 Two video pictures of an *Aplysia* growth cone; (b) was taken 7 min after (a).

☐ What do the pictures in Figure 2.14 remind you of?

■ They should remind you of the picture of the *Amoeba* in Book 2, Section 2.2.3, without its nucleus.

Films of living growth cones in tissue culture (see Box 2.2) show that they move in the same way as *Amoeba*. The spiky protrusions extending from the growth cone are called *filopodia* (singular *filopodium*). Some appear to extend and then attach to whatever they are growing through (referred to as the *substrate*), whereas others are retracted. The cytoplasm in the central region then flows into the 'web' between the filopodia. This moves the growth cone forwards as the filopodia extend further. The forward movement is probably the result of both a pulling force from the filopodia (they contain contractile filaments) and a pushing force from the flow of material into the growth cone. As the growth cone moves forward, it produces the axon behind. The rate of growth is about 10–$40\,\mu\text{m h}^{-1}$.

Box 2.2   Tissue culture

Embryonic neurons, like many other types of cell, will survive and grow outside the body, provided they are bathed in an appropriate fluid at the right temperature. The technique of growing cells outside the body is called *tissue culture* and was pioneered in 1910 by an American, Ross Harrison. He wished to prove that axons really were long extensions of single neurons and that they could extend to reach targets some distance away. He placed some tissue from the spinal cord of a tadpole in a dish and made sure that the fluid bathing the tissue contained oxygen and had a similar composition to the extracellular fluid of the tadpole. Under these, rather strict, conditions he saw that the neurons from within the spinal cord sprouted long extensions, which then grew into longer and longer threads. The processes growing out of neurons in culture could be either dendrites or axons, and collectively are called **neurites**. The similarity of the growing tips of neurites to growth cones seen in embryos suggests that it is safe to assume that their behaviour resembles that of normal axonal (or dendritic) growth cones.

Tissue culture has the advantage that the behaviour of growing axons can be analysed in detail and the environment can be controlled so that possible factors that influence the direction of growth—*guidance cues*—can be investigated.

☐   What is the disadvantage of tissue culture?

■   The cells grow in a very artificial environment. Although the conditions can be controlled in tissue culture, they can never match those found in the normal developing embryo. What happens to neurons in a dish—for instance how they respond to possible guidance cues—may not reflect how neurons respond in their normal, much more complex, environment.

The stimuli (cues) that guide axons to their destinations within the nervous system and to targets in the rest of the body have intrigued developmental neurobiologists ever since it was realized that these extensions of neurons often connect cells over great distances. So what are these cues and how are axons guided? How do axons 'know' where to grow? These questions are addressed in subsequent sections. First, the next section deals with general issues of axon growth.

2.4.2 A pattern of axons

The pattern of axons making up the peripheral nerves in a growing limb is very striking. Figure 2.15 shows a dorsal view of the pattern of nerves (stained with a silver-based dye) in the left and right wings of a chick at E8 (chicks hatch at E21).

☐ Compare the pattern in the two limbs and make a general statement about what you see.

■ The patterns are very similar: the pattern in one limb is virtually the mirror image of the pattern in the other limb.

Not only is the pattern similar between the limbs of the same chick, it is also very similar between different chicks.

☐ What does the similarity in chick wing pattern suggest about the growth of axons within the chick limb?

■ The similarity suggests that the growth of axons is controlled in some way.

Thus, axons do not grow haphazardly. Rather, the path that an axon takes, and hence the pattern of axons in the limb, is very precisely controlled. In principle, the control could be exerted in two very broad ways: (a) as an intrinsic property of individual neurons and (b) as a result of guides or cues in the limb, which interact with and direct the growing axon.

In fact, the path of an individual axon is not an intrinsic property of the neuron. In other words a neuron does not grow p millimetres, turn x degrees, grow q millimetres, etc. This has been shown by moving a limb bud from an E3 chick embryo and transplanting it onto a different site in a host embryo, before the motor and sensory axons had grown out into it. These transplanted limb buds heal well in the host, and continue to develop into normal limbs. They become innervated by axons that would not normally innervate a limb. The pattern of axons in such transplanted limbs does not reflect the pattern that those axons would normally make; the pattern is typical for the limb and is remarkably normal. The growing axons thus seem to be following general pathways laid down in the limb.

Two further pieces of evidence support the conclusion that there are guidance cues of some sort in the limb. The first piece of evidence addresses the question of whether the target influences the path of the growing axon. In the limb bud this would mean that the motor axons follow a path dictated by the muscles they are growing towards. Each muscle is made up of cells that migrate from particular somites (see Figure 2.4, p. 17). The somites can be removed before migration by careful surgery early in development so that the limb develops without muscles. In the early pattern of nerves in such muscleless limbs, the muscle nerve branches are present in their normal positions. This suggests, for the initial paths at least, that guidance cues are not provided by the muscles.

The second piece of evidence relates the pattern of axons in the limb to the types of tissue through which the axons grow. Essentially, axons do not penetrate areas of the developing limb that will later form cartilage and bone. These areas of dense tissue force the axons to grow through other less-dense areas where there are gaps between the cells. The environment in the gaps is known as the **extracellular**

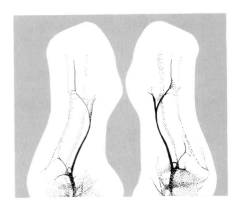

1 mm

Figure 2.15 Drawings of left and right wings of a chick embryo at E8, showing the pattern of nerves viewed from the dorsal side.

matrix. It consists of proteins and other substances that are produced by the cells, and also the outer membranes of the cells themselves.

The precision of the growth of the axons, and the evidence that there is a pattern of pathways that axons follow, show that the environment of the growth cone plays a very important role in guiding growing axons. The following sections illustrate how growing axons are guided by factors in the extracellular matrix and on the surface membranes of cells they encounter.

2.4.3   Finding a way—pathways

The growth cone moves forward by temporarily sticking its extending filopodia to the substrate (see Section 2.4.1 above). It follows that, if the filopodia cannot stick to the substrate, then growth over that substrate will be very slow, or absent; conversely, growth over a substrate to which the filopodia stick well will be faster. The advance of the growth cone has been likened to that of a climber: the climber's progress up a rock face is aided by finding good holds.

Paul Letourneau (1975) demonstrated this phenomenon by growing neurons in tissue culture. He used two substrates, palladium and the amino acid polymer polyornithine. By gently squirting fluid at the filopodia, he found that very little force was required to dislodge those growing on palladium, whereas much more force was required to dislodge filopodia growing on a substrate of polyornithine.

☐   If the substrate on which neurites (the collective term for axons and dendrites; see Box 2.2) are growing consists of squares of palladium separated by alleys of polyornithine, where would you expect to find the neurites?

■   You would expect to find the neurites in the alleys, because that is where the filopodia can stick and make progress.

Just such an experiment is illustrated in Figure 2.16, which shows neurites from dorsal root ganglion cells growing on a culture dish coated with palladium metal in a pattern of squares. The neurites in Figure 2.16 are confined to the alleys coated with polyornithine.

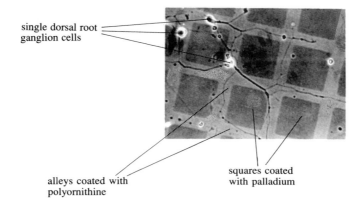

single dorsal root
ganglion cells

alleys coated with
polyornithine

squares coated
with palladium

Figure 2.16 The pattern of growth of neurites from dorsal root ganglion cells on a pattern of substrate in which polyornithine forms alleys between squares of palladium. The neurons were taken from E8 chick embryos.

So here in a culture dish can be seen a clear example of how axon pathways might be produced by axons following *adhesion* cues. Do such cues exist in nature and if so, what are they? An important technique for identifying substances that may be involved in axon guidance is immunohistochemistry (Box 2.3). This technique provides a picture of the distribution of particular substances within the tissue.

Box 2.3 Immunohistochemistry

Immunohistochemistry is the term for any technique for locating a substance in a tissue preparation using an antibody specific for that substance. As you learnt in Book 2, Chapter 5, the immune system of vertebrates reacts to the presence of pathogens (antigens) by producing substances called antibodies, which bind specifically to the antigen, leading to the neutralization of the antigen by white blood cells (Figure 2.17).

Antibodies will only combine chemically with the specific antigens that they match. It is this property that is used to determine whether a particular tissue sample contains a specific antigen. The sample is bathed in fluid containing the antibody, and is then washed. Any antibody remaining after washing must be attached to antigen present in the tissue (Figure 2.17). The use of antibodies to detect the presence of certain antigens in tissue samples (histological specimens) requires that the antibodies are made visible. This is done by attaching a 'marker' molecule to the antibody which can be seen under the microscope or in some other way. Commonly used markers are the enzyme horseradish peroxidase, HRP, a fluorescent molecule, or a radioactive molecule.

When an animal is immunized, it will produce a variety of antibodies to the antigen (*polyclonal antibodies*). There are methods for producing antibodies of only one type by culturing spleen cells from the immunized animals. These pure antibodies are called *monoclonal antibodies*. Monoclonal antibodies become important when detailed analysis of the location of specific antigens and their identity is required.

Immunohistochemical techniques have been used to identify and observe the distribution of extracellular molecules important in development as well as to investigate neuronal projections and neurotransmitters. As antibody binding often blocks the function of such extracellular molecules, these techniques can also be used to investigate the roles of particular molecules during development.

Adhesion cues do exist, and one example of such a cue is the protein laminin. Laminin is an important component of the extracellular matrix material that surrounds all peripheral nerves, muscles and also the brain in vertebrates. (It is also present in invertebrates.)

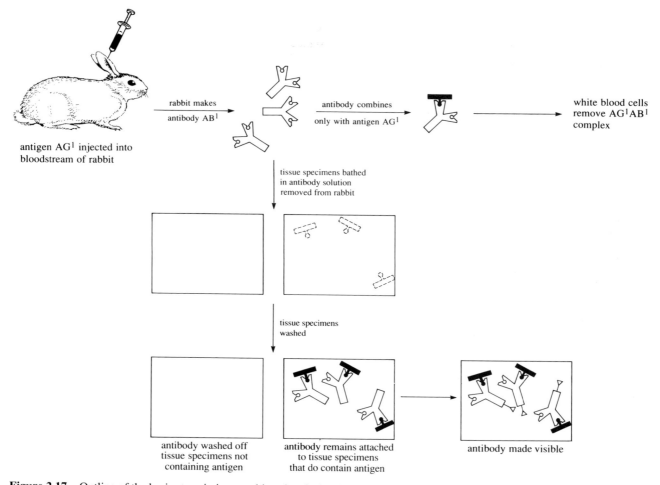

Figure 2.17   Outline of the basic steps in immunohistochemical staining techniques. AB[1] is the antibody specific to the antigen AG[1].

Any substance that guides axons by adhesion should meet three criteria:

1   It should be present along the pathways of growing axons.

2   It should promote neurite outgrowth in tissue culture.

3   The axons themselves should have receptors in their membranes which can bind to the substance.

Laminin meets all three criteria: it is present, for example, in the chick optic nerve and tracts when the retinal axons are growing to the optic tectum (for location, see figure of a bird brain on fold-out page at the end of this book); in tissue culture, neurite outgrowth from dorsal root ganglion cells is much greater on a substrate coated with laminin than on other purified components of the extracellular matrix; and the axons bind to the laminin via cell-surface receptors. The cell-surface receptors, of which there are many examples, are composed of substances called *glycoproteins*. Some cell-surface receptors, called cell-adhesion molecules, bind only to their own kind. The best-known cell-adhesion molecule is called *neural cell-adhesion molecule* (NCAM).

There are thus two types of cell adhesion. In one, cell-surface receptors bind to a different molecule; in the other, cell-surface receptors bind to an identical glycoprotein.

Three pieces of evidence point to the importance of NCAM in development. Firstly, the distribution of NCAM in the developing chick optic tract almost exactly overlaps that of laminin. Secondly, normal axon growth in the retina is disrupted after treatment with antibodies to NCAM. And thirdly, in the developing chick limb, NCAM is present along the growing nerves (rather than ahead of them) and also on developing muscle cells. It therefore plays a role in establishing the pattern of nerves within the limb. All this evidence suggests that NCAM helps to define pathways of adhesion and to bind axons together via NCAM linkages, thus promoting growth of axons along particular pathways.

The growth of axons can also be affected by extracellular substances that repel axons. In Section 2.3.5 the migration of neural crest cells through the rostral halves of the somites was described. After the neural crest cells have migrated, the outgrowth of axons from the spinal cord is also confined to the rostral halves of each somite. Immunohistochemical staining (see Box 2.3) shows that both laminin and NCAM are present in the areas traversed by neural crest cells and axons. The axons never penetrate the caudal halves of each somite. The response of growth cones to cells from rostral and caudal halves of somites in tissue culture shows that they avoid contacting cells from the caudal half. An extract made from the caudal half of somite tissue causes growth cones to collapse in tissue culture. The confinement of axons (and neural crest cells) to the rostral half of each somite is due to the presence of adhesive cues such as NCAM in the rostral half, and also to the fact that the cells in the caudal half repel rather than attract axons.

The presence of adhesive and repulsive substances along the pathways taken by growing axons suggests that the pathway followed by axons is not defined by the presence of a single chemical cue. The fact that axons of many different types of neuron respond to these cues in similar fashion also suggests that these substances do not provide a sufficient explanation for the exquisite detail of some patterns of axonal projection (for example the topographic projections in the visual system); other cues must also operate.

2.4.4 Finding a way—guide-posts

In the 1960s a chance observation that the horseradish peroxidase (HRP) histochemical methods used for tracing nerve connections (Book 2, Box 8.1) stained complete cells in the developing insect nervous system, started off a long and productive line of research into axon guidance in these animals.

The sensory neurons in a grasshopper (*Schistocerca nitens*) limb originate from cells in the periphery, their axons growing towards the ventral nerve cord. Look at the three drawings on the left in Figure 2.18. The top drawing shows the initial outgrowth of the axons of two sensory neurons that lie near the tip of the limb. They grow towards particular non-neuronal cells numbered 1 and 2 (middle drawing). At cell 2 there is a change of direction towards non-neuronal cell 3, after which there is a further change of direction (bottom drawing). The axons reach their target cells in the ventral nerve cord about 100–150 μm away.

When the axons contact cells 2 and 3, they change their direction of growth. The contact made between the axons and the cells is very close. The axons actually seem to push processes into the cells. The cells that the axons contact appear to be acting like 'stepping stones' or 'guide-posts', guiding the axons to the central nervous system. Support for this hypothesis is seen from the drawings on the right of Figure 2.18. Cell 3 was destroyed before the outgrowing axons had reached it. The growing axons begin their growth normally, reaching cells 1 and 2. However, the axons then spread out in all directions, covering a much greater area of the limb, and, within the same time-scale as the bottom left drawing, do not reach the ventral nerve cord. (In a number of cases the axons do eventually reach the ventral nerve cord.) Cell 3 thus seems to act like a guide-post, with the growing axons changing direction when they meet it.

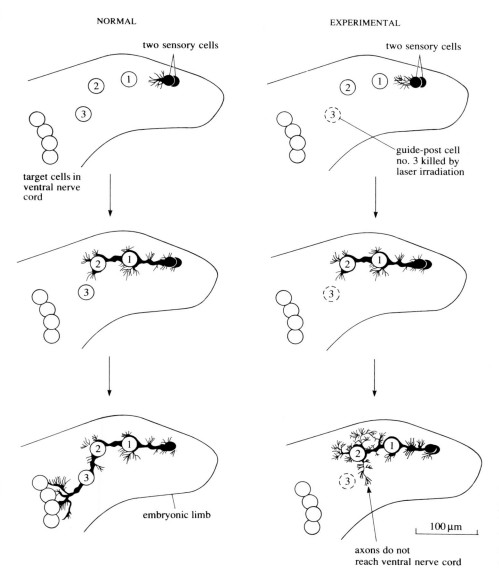

Figure 2.18   Guide-posts in the grasshopper limb. The three drawings on the left show the normal pathway to the CNS taken by axons of two sensory neurons. Also shown are three non-neuronal guide-post cells, 1, 2 and 3. The three drawings on the right show the effect on axon growth when cell 3 is destroyed by laser irradiation. Growth to cells 1 and 2 is normal, but the appropriate changes of direction after cell 2 are not made.

This example also illustrates another feature of development. The initial pathway shown in Figure 2.18 was established by axons from two sensory cells. However, many neurons that develop later will send their axons along the same path. (Hence the neurons that establish the path are sometimes referred to as *pioneer* cells.) Insect axons, like those of vertebrates, have adhesion molecules on their surface similar to the neural cell adhesion molecules (Section 2.4.3), which serve to bind the separate axons into a tight bundle. The presence of such adhesion molecules makes the task of the axons following the pioneer axons that much easier.

In the experiment cited above, the importance of guide-post cells in guiding pioneer axons in the insect limb was demonstrated by selectively destroying the guide-post cells. Another way of investigating axon guidance cues is to alter the relationship between the axon and its target surgically, and then to see if the axons still manage to reach their correct destinations. Such experiments have shown that cues also exist to guide motor axons to muscles in the chick limb.

Each muscle in the vertebrate limb is innervated by a cluster of motor neurons (called its *motor pool*), situated in a characteristic position within the ventral spinal cord. The arrangement of the motor pools of different muscles is related to the position of the muscles in the limb in a very straightforward way, summarized in Figure 2.19a. Notice that the relative positions of the motor pools are the same as the relative positions of the muscles A, B and C that they innervate. This straight-forward mapping is achieved despite the fact that the motor axons are in different segmental nerves that come together and then separate to innervate each of the muscles. Figure 2.19b shows what happens if a portion of the spinal cord is reversed end for end before the axons grow out.

The two important features of this experiment are:

1   The axons make contact with their appropriate target even after the operation.

2   Groups of axons, in taking a different path to their targets, cross over one another.

When much longer segments of spinal cord were reversed, displacing the axons further from their normal routes, many motor neurons did not reach their normal target muscles. The cues that guide the axons to reach specific muscles therefore seem only to be present close to their normal routes.

The limb transplantation experiments described in Section 2.4.2 point to the existence of pathways that axons grow along preferentially. The spinal cord reversal experiments show that there are also local cues that guide particular axons towards particular targets. In the chick limb there are no obvious pioneer cells, and no cells that seem to be acting as guide-posts. At the time of writing (1992) the nature of the specific cues to which the motor axons are responding is unknown.

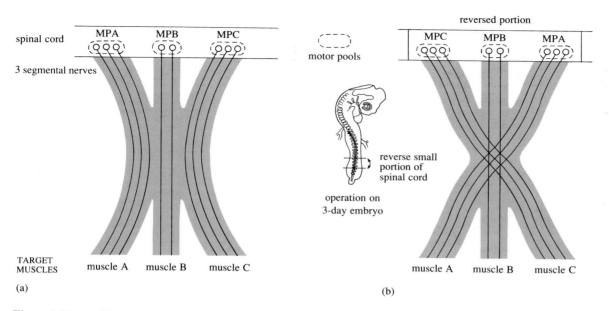

Figure 2.19   (a) Diagram showing the arrangement of three motor pools (MPA, MPB and MPC), their output via three segmental nerves and their target muscles (muscle A, muscle B, muscle C) in the chick limb. The arrangement of the pools reflects the distribution of their targets in the limb. (b) Diagram showing what happens after reversal of a segment of spinal cord (operation at E3). Displaced motor neurons still innervate target muscles appropriate for their original positions in the cord. The motor axons do not grow in parallel but have to cross over one another to reach their target muscles.

2.4.5   Finding a way—guidance at a distance

So far, this section has described how axon pathways might be defined by contact-mediated interactions between the axons and their environment. Such guidance is often referred to as **chemotactic guidance**. In 1910 Cajal suggested that axons could also be attracted to their targets by a chemical attractant acting at a distance, much as a male moth flies towards the source of pheromone released by the female moth (Book 1, Section 2.7.1). This kind of guidance is called **chemotropic guidance**. The idea is that the target releases the attractant and the axon grows towards the source of the attractant and so towards the target. In practice, it has proved difficult to distinguish between chemotactic guidance cues, where contact is required between the growth cone and other cells, and chemotropic guidance cues, which do not require contact.

Andrew Lumsden and Alun Davies in London (1986) demonstrated chemotropic guidance in the growth of axons from the sensory ganglion—the trigeminal ganglion—that innervates the whiskers on the muzzle of the mouse (*Mus musculus*). The whiskers develop in a particular region of the face called the *maxillary region*. At E9, axons start growing from the trigeminal ganglion towards the maxillary region some 20 µm away. At E11, the axons have reached the maxillary region and by E12 they start to innervate the developing whiskers.

The trigeminal ganglion and the maxillary region can easily be removed from the embryo and placed in tissue culture. Lumsden and Davies did this, placing the maxillary region on one side of the ganglion and a limb bud (which the ganglion does not normally innervate) on the other. Each target was at a distance of 10–20 μm from the ganglion. They then recorded which way the neurites grew out from the ganglion.

☐ What would you expect to happen if the maxillary region produces an attractant?

■ The neurites from the trigeminal ganglion should grow towards the maxillary region and not towards the limb bud.

☐ Does the outgrowth of fibres shown in Figure 2.20a support the chemotropic guidance hypothesis?

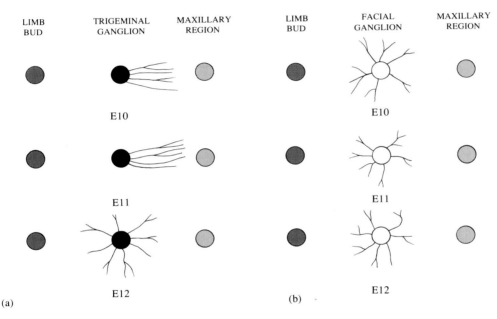

LIMB BUD · TRIGEMINAL GANGLION · MAXILLARY REGION · LIMB BUD · FACIAL GANGLION · MAXILLARY REGION

E10 · E11 · E12

(a) · (b)

Figure 2.20 Drawings of neurite outgrowth from ganglia taken from E10, E11 and E12 mice embryos. To the left of the ganglion is limb bud tissue, to the right of the ganglion is tissue from the maxillary region: (a) trigeminal ganglia; (b) facial nerve ganglia.

■ Yes it does. Neurites from the trigeminal ganglia of E10 and E11 embryos grew towards the maxillary tissue and not towards the limb bud.

Now look at Figure 2.20b, which shows the results of a similar tissue culture experiment using facial nerve ganglion instead of trigeminal ganglion.

☐ The neurites from the facial nerve ganglion grow randomly. What does this suggest?

■ The random growth from this ganglion suggests that, although able to grow, these neurites were unable to respond to the attractant released by the maxillary region.

The facial nerve ganglion is actually next to the trigeminal ganglion; they both innervate parts of the face, but only the trigeminal ganglion innervates the maxillary region. The chemotropic cues released by the maxillary region therefore act specifically on those axons from the trigeminal ganglion that will innervate it.

There was random growth from both the trigeminal and facial nerve ganglia taken from E12 embryos.

☐   What does the random growth of E12 trigeminal neurites suggest?

■   The random growth of the E12 trigeminal neurites suggests one of two things. Either the maxillary region stops producing the attractant at E12 (that is, there is no chemotropic agent present), or the E12 trigeminal neurites are no longer able to respond to the attractant. The implications of this observation are discussed in Section 2.4.7.

There is further evidence that the target releases a specific chemotropic agent. If the maxillary region is grown in culture in a gel for 48 h, then removed and the trigeminal ganglion placed in the gel near where the maxillary region had been, the outgrowth of neurites from the trigeminal ganglion is directed towards the position where the target (the maxillary region) had been.

☐   If the gel is rinsed in saline solution before adding the ganglion, outgrowth is no longer directed. What does this suggest?

■   This experiment suggests that whatever was responsible for directing the growth was washed away by the saline solution.

All these tests indicate that the chemotropic agent (called 'maxfactor'!) must be a diffusible molecule, which specifically attracts the growing trigeminal axons.

2.4.6   Inhibiting growth

So far in this section, only those factors that affect the direction in which the axon grows have been considered. However, it is perfectly obvious that at some point the axon, and the dendrites, have to stop growing. For the axon this appears to be largely dependent on contact with the target (see Chapter 3). The growth of dendrites appears to be influenced by neurotransmitters, however.

In the 1980s Stanley Kater in Colorado noticed that neurites from cultured *Aplysia californica* neurons stopped growing when they came close to one another. Further experiments revealed that the signal in this case seemed to be the presence of one or more neurotransmitter. More recently he has looked at this phenomenon in hippocampal pyramidal cells in culture. The advantage of using hippocampal cells (taken from E16 rat embryos) is that, in culture, it is possible to identify both the axon and the dendrites of these cells. Also it is known that glutamate is one of the more important neurotransmitters acting on these cells. Figure 2.21 (overleaf) shows the effect of treating growing hippocampal cells with glutamate.

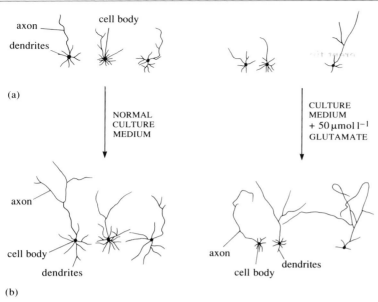

Figure 2.21 Effects of glutamate on outgrowth from hippocampal cells. (a) Cells from E16 rat embryos after being grown in culture for 2 days. (b) Between days 3 and 5 in culture the cells were exposed either to normal culture medium or to the same culture medium to which 50 μmol l^{-1} (μmol l^{-1} is a unit of concentration) glutamate had been added.

☐ What effect does glutamate exposure have on dendritic and axonal growth of these cells?

■ The transmitter seems to stunt the growth of the dendrites, but does not affect the axons.

Using a micropipette (a very fine glass tube) it is possible to apply glutamate to specific dendrites. Such application affects the growth only of those specific dendrites and no others. Thus, glutamate has a very localized effect on growing dendrites. If this result can be generalized to other neurotransmitters, then it is possible that cumulative effects of different neurotransmitters released by growing axons onto neurons may be involved in determining the dendritic tree, that is, the form and branching of dendrites.

2.4.7 All change

Figure 2.20 showed the growth of neurites from the trigeminal ganglion towards the maxillary region. The neurites grew towards the maxillary region at ages E10 and E11, but not E12. The interpretation of this observation was either that by age E12 the maxillary region had stopped producing the chemotropic agent, or that by E12 the trigeminal neurites had stopped responding to the agent (and indeed both interpretations may be true). Both these interpretations have two very important implications. First, in the normal course of events, the time when trigeminal axons are growing must coincide precisely with the time when the maxillary region is producing (or the trigeminal axon is responding to) its chemotropic agent. The second implication is that factors (or responses) that affect neuronal growth are not present all the time. There are certain times when some factors or responses are present and other times when other factors or responses are present. In other words, the developmental map is different at different times. If a growing axon does not reach a particular place at a particular time, the guidance cue it needs in order to grow correctly may have already disappeared. Alternatively, the axon itself may have changed, so that it no longer responds to the chemotropic agent. Further work is necessary before it is possible to establish which of the above interpretations is correct.

Two further pieces of evidence support the idea of a developmental map that changes with age. The first piece of evidence involves laminin. In Section 2.4.3 you learnt that axons grow along laminin paths. However, neurites from the chick retinal ganglion cells grow best on laminin if they come from embryos between E6 and E11. Younger and older retinal ganglion cells do not show such a response to laminin. The E6–E11 time period exactly matches the time when retinal axons are growing to the tectum. So, as the retinal cells differentiate, their cell-surface receptors also change. Furthermore, the laminin disappears from the retino-tectal pathway as soon as the axons have reached the tectum.

☐   What does this suggest?

■   It suggests that the cells along the pathway to be followed by the growing retinal axons only produce laminin during the time the axons are growing.

The second piece of evidence is that neural cell adhesion molecule (NCAM) that has been purified from early embryos binds axons less strongly than that purified from adult animals. It has been suggested that the less-adhesive variety reflects the greater mobility required by axons during growth. The increased adhesion between axons in some pathways might prevent other growing axons from invading them. Thus, variation in NCAM over time might play a role in the separation of pathways in the brain.

2.4.8   Summary of axon guidance

The growth of axons to their targets involves a complex interplay of events in many types of cell. The interaction between axons and their environment is a two-way process. While growing, axons are affected by their environment, so that they in turn will affect the cells through which they are growing. Such interactions are not confined to the growth cone. The parent cell body responds to signals from its immediate environment (such as arriving axons) and also to signals absorbed by the growing axon tip and transported back to the cell body.

It has long been realized that there are far too many precise connections in the nervous system to be accounted for by individually specifying each neuron and each target with its own recognition molecule. The timing and appearance of guidance molecules demonstrates clearly that the same molecule or its receptor can be used to fashion different pathways at different times in development. There is also evidence that the release of neurotransmitters can affect the growth of nerve processes.

The many experiments involving transplantation and interference have shown that axons seem to use multiple cues to reach their destinations. Neurobiologists are beginning to learn about general axonal pathways shaped by relative adhesion or repulsion, but as yet the molecular nature of possible detailed cues that axons might use when close to particular targets is unknown.

Summary of Chapter 2

This chapter began by outlining the main phases of development, including cell migration, differentiation and growth. These phases were then examined with specific reference to neurons.

The nervous system starts off as a simple hollow nerve cord made up of ectodermal cells (the neural tube). After a neuron has been born and has migrated to its final position in the nervous system, its development depends both on its lineage and also on its local environment. Apart from differentiating, neurons also have to grow axons and dendrites, and contact the correct target. The principal ways in which axons are guided with precision to their targets were examined. The various cues, including guide-post cells, pathways, and chemotactic and chemotropic guidance, operate in both space (different cues being present in different places within the embryo) and time (cues may be present or detected at some times and not others) to guide axons to their targets.

Once the axons reach their targets, further complex interactions occur, which are the topics of the next chapter.

Objectives for Chapter 2

When you have completed this chapter, you should be able to:

2.1 Define and use, or recognize definitions and applications of, each of the terms printed in **bold** in the text. (*Question 2.4*)

2.2 Describe the early stages of the neural development of the embryo. (*Questions 2.1, 2.3 and 2.5*)

2.3 Discuss factors that influence the differentiation of cells. (*Questions 2.2, 2.3 and 2.5*)

2.4 Explain how transplantation experiments demonstrate the importance of both time and place in development. (*Question 2.4*)

2.5 Describe how axons grow, and discuss the role of adhesion in growth. (*Question 2.6*)

2.6 Discuss evidence that suggests that axons respond to a variety of cues when growing towards their targets. (*Questions 2.6, 2.7 and 2.8*)

Questions for Chapter 2

Question 2.1 (*Objective 2.2*)
Why does the grafting of ectoderm from the dorsal blastopore region to the belly region of a host embryo only result in a double embryo if the operation is performed at a particular stage of development?

Question 2.2 (*Objective 2.3*)
Why is it easier to investigate cell lineage in simple animals? What methods are available for analysing lineage in vertebrates?

Question 2.3 (*Objectives 2.2 and 2.3*)
What is the evidence that neurons migrate away from the place where they were born, in (a) the cerebral cortex and (b) the peripheral nervous system?

Question 2.4 (*Objectives 2.1 and 2.4*)
In general terms, how does induction occur?

Question 2.5 (*Objectives 2.2 and 2.3*)
What are the destinations of early- and late-born neurons in the cerebral cortex, and how has the study of the reeler mutant mouse helped in understanding the role of birth date and cell position in neuronal differentiation?

Question 2.6 (*Objectives 2.5 and 2.6*)
Comment on the statement 'all growing axons make straight for their targets without hesitation or deviation'.

Question 2.7 (*Objective 2.6*)
What is the difference between chemotactic and chemotropic factors in the way they affect axon growth?

Question 2.8 (*Objective 2.6*)
Why is it unrealistic to think that a single factor could guide axons to their destinations?

References

Letourneau, P. C. (1975), Cell to substratum adhesion and guidance of axonal elongation, *Developmental Biology*, **44**, pp. 92–101.

Lumsden, A. G. S., and Davies, A. M. (1986). Chemotropic effect of specific target epithelium in the development of the mammalian nervous system, *Nature*, **323**, pp. 538–539.

Further reading

Alberts, B., Bray, D., Lewis, J., Raff, M., Roberts, K. and Watson, J. D. (1994). *Molecular Biology of the Cell*, 3rd edn, Chapter 21: Cellular Mechanisms of Development (section on neural development), pp.1119–1137, Garland Publishing.

Brown, M. C, Hopkins, W. G. and Keynes, R. (1991). *Essentials of Neural Development*, Parts 1–3, Cambridge University Press.

Parnavelas, J. G., Stern, C. D. and Stirling, R. V. (eds) (1988). *The Making of the Nervous System*. Part 3, Molecules and guidance pathways; Part 4, Sharpening the pattern, Oxford University Press.

Purves, D. and Lichtman, J. W. (1985). *Principles of Neural Development*, Sinauer Associates.

CHAPTER 3
NEURON–TARGET INTERACTIONS IN DEVELOPING AND ADULT ANIMALS

This chapter falls into two parts. The first part (Sections 3.1–3.4) continues the story of the development of the nervous system from the point where the neurons contact their targets. Once neurons have reached their targets, they start to form synapses, a process called **synaptogenesis**. This process is the subject of Section 3.1. Concurrent with synaptogenesis, two other major developmental processes are occurring. One is the loss of a very large number of neurons. Why there should be such a loss, and the ways in which neurons are selected to survive, form the subject matter of Sections 3.2 and 3.3. The other major process is the refinement of synapses after they have been made. There is now a great deal of information available about how axons affect their targets and how the targets in turn affect the innervating neurons. The continued maturation of both the neuron and its target (be it another neuron, a sense organ, a gland or a muscle) depends critically on the contact between them. Such interactions play a vital role in establishing precise connections in the nervous system, and are considered in Section 3.4.

The second part of this chapter (Sections 3.5 and 3.6) deals with the medically important area of the response of the mature nervous system to injury. Some, but not all, parts of the mature nervous system retain the ability to grow. Sections 3.5 and 3.6 explore this ability, and the similarities between growth in the mature nervous system and growth in the developing system.

A number of the experiments reported in this chapter raise difficult questions about the ethics of animal experimentation. They are reported here because they provide important insights into developmental processes with possible medical implications.

The chapter begins at the point where the axon reaches its target.

3.1 Establishing contact, forming synapses

When axons reach their targets, they start to form synaptic contacts and stop growing. Many studies of the events occurring during synapse formation in muscle cells have been made in tissue culture, where axons growing out from slices of spinal cord innervate muscle cells placed nearby. In this situation it is possible to monitor the exact time when axons contact a muscle cell, and to record any ensuing electrical events that occur in the muscle cell.

Within 20 minutes of a motor neuron axon contacting a muscle cell, there is a marked increase in the rate of release of neurotransmitter from the motor axon.

☐   What would be good evidence for this increase?

■   An increase in both the size and frequency of postsynaptic potentials recorded from the muscle cell (Book 2, Section 4.2.4).

At this time the axon becomes quite firmly attached to the muscle cell, despite the presence of the synaptic cleft. During the next two hours the postsynaptic membrane thickens and acetylcholine receptors concentrate beneath the contact point.

☐   How would you demonstrate the presence of the acetylcholine receptors?

■   The presence and distribution of the acetylcholine receptors can be revealed by antibodies that recognize the receptor molecule (Box 2.3). Alternatively, they could be labelled with something that binds to the receptor, for example radiolabelled α-bungarotoxin (Book 2, Section 4.2.5)

A little later, vesicles begin to cluster in the presynaptic terminal, and the postsynaptic membrane begins to differentiate to form the characteristic complex structure of the neuromuscular junction (Book 2, Section 2.4.2).

From this description you can see that the formation of synaptic contacts involves changes in both the presynaptic neuron and the postsynaptic cell. This contact is a very effective communication channel between neurons and their targets. In addition to the release of transmitter by the presynaptic cell, other signalling substances are released and received by both cells. Thus, information, in the form of chemical messages, passes between the two cells in both directions.

The continued differentiation and survival of both presynaptic and postsynaptic cells depends on the maintenance of the contact between the two. If muscle cells fail to receive motor innervation they do not differentiate and eventually die. The development and maintenance of many sensory receptors (for example taste buds in the tongue (Book 3, Section 1.2.4) or muscle spindles in muscle) likewise depend on intact, functional innervation. In the central nervous system the removal of a sense organ during development leads to a decrease in the growth and differentiation of their target neurons, and sometimes to the death of these neurons. Also, the removal of a target can lead to the death of those neurons that would normally project to that target. This dependence on contact is much more common in developing systems than in mature animals: cutting motor axons or removing muscles results in loss of motor neurons during development but not in mature animals.

3.2   Selective neuronal survival in the nervous system

Selective neuronal survival is an important process in the development of the detailed matching between neurons and their targets. In most parts of the vertebrate nervous system the period of formation of synaptic connections coincides with a massive loss (20–80%) of neurons. At first sight this loss of cells might seem to be rather wasteful, but it has two important advantages. One is that it provides a mechanism for matching the number of neurons to the size of the target, allowing for variations in, for example, (the sizes of muscles.) Another is that it eliminates those neurons that have reached inappropriate targets.) In the following sections, evidence for these two processes is presented.

3.2.1 Neuronal survival—the influence of the target

Figure 3.1 is a graph showing the numbers of motor neurons in a part of the spinal cord (the lumbar ventral horn) in the chick at various times during development.

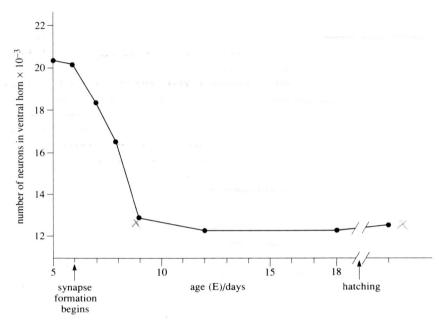

Figure 3.1 Graph showing the number of motor neurons in the lumbar ventral horn of the chick at various stages of development of the chick embryo. Hatching takes place at E21.

☐ What do you conclude from Figure 3.1 about what happens to neurons during development? *of chick in egg.*

■ The graph shows that many neurons disappear (they actually die) during development, mostly between E6 and E9.

The timing of this period of cell death is concurrent with the time when the motor neuron axons are forming synapses with muscles in the limb. This is not just a fortunate coincidence. Neuronal survival depends on the successful completion of two stages. The first stage is innervation; neurons that do not innervate an appropriate target die. However, innervation itself does not guarantee survival; it merely allows neurons to enter the second stage. In the second stage, the only neurons to survive are those that have made a sufficient number of functional contacts (that is, working synapses) with their target. Formation of just a few synapses does not protect a neuron from death. So how are the neurons that subsequently survive selected from the rest? The following observations suggest an answer.

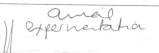

In the chicken, if an extra developing limb is grafted close to the normal limb, before the period of cell death, the number of motor neurons or dorsal root ganglion cells that survive is increased. Conversely, if limb muscles are removed, then the number of motor neurons that survive is smaller. Survival of neurons appears to be proportional to the amount of limb tissue. One way to explain this relationship is by invoking some kind of 'factor' produced by the limb tissue. Those neurons that acquire sufficient amounts of the factor survive; the others die. The neurons, in effect, compete for a 'survival factor' produced by the target cells. The selective survival of neurons in other regions of the nervous system (for example the retina and the sympathetic nervous system) also seems to involve competition for a survival factor produced by the target cells.

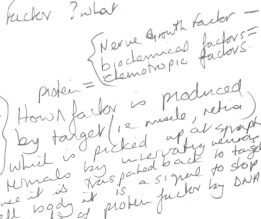

How might such a factor influence neuronal survival? One idea is that the target releases the factor, which is then picked up at the synaptic terminals by the innervating neurons. It is then transported back to the cell body, where it acts as a signal to the DNA to produce or to stop producing particular proteins, and thus affects the functioning and survival of the cell.

3.2.2 Biochemical factors that mediate neuronal survival

There is much additional evidence showing that many kinds of neuron must obtain a survival factor from their targets. Biochemical factors that promote cell survival are known as **chemotrophic factors** (not to be confused with chemotropic factors—Section 2.4.5). Figure 3.2b shows that removal of the wing bud in a chick embryo results in a decrease in the size of the adjacent dorsal root ganglia (such as A in Figure 3.2) containing sensory neurons, whereas grafting an extra wing bud leads to the innervation of that wing bud and an increase in the size of the ganglia (Figure 3.2c).

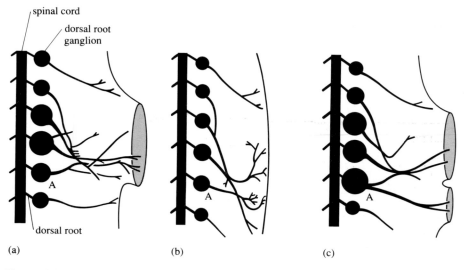

(a)   (b)   (c)

Figure 3.2   The effects of removing a wing bud and grafting a wing bud on the size of nearby dorsal root ganglia: (a) normal situation; (b) removal of a wing bud results in a reduction in the size of ganglia such as A, whose axons had innervated it; (c) ganglion A is larger after an extra wing bud has been grafted close to it.

This is a clear case in which a target tissue seems to be releasing a factor that promotes the survival of neurons in the dorsal root. A similar effect has been found in ganglia in the sympathetic nervous system (Book 2, Section 8.5). The story of how this factor was purified and identified (Levi-Montalcini, 1975)—for which Rita Levi-Montalcini received the Nobel Prize for Physiology or Medicine in 1986—is interesting, emphasising the importance of chance in biological investigation. In an early attempt to identify the factor responsible, a fast-growing tumour was transplanted to a position near the spine in a chick embryo. Axons grew into the tumour, and the size of both the dorsal root ganglia and the sympathetic ganglia next to the tumour were greatly enlarged, although there was no effect on neurons within the spinal cord. (Dorsal root and sympathetic ganglia lie just outside the spinal cord; see Figure 3.2.) In 1953, Levi-Montalcini and Viktor Hamburger tested whether actual contact with the tumour was important by growing the tumour on the membrane that surrounds the embryo in the egg, thus eliminating any contact between the tumour and the nervous system. This membrane has a very good blood supply, so any substances produced by the tumour would be carried into the bloodstream of the embryo. The result was quite dramatic; there was a large increase in the size of all dorsal root and sympathetic ganglia throughout the embryo. An agent must therefore have been transported from the tumour to the ganglia via the bloodstream. Levi-Montalcini and Hamburger called this agent **nerve growth factor (NGF)**.

The next step in the identification of NGF involved tissue culture. Dorsal root ganglia were removed from chicken embryos at E6–8 days and grown in tissue culture. The survival of neurons in the ganglia and the outgrowth of neurites can be easily monitored in such cultures, allowing the effects of adding various substances extracted from the tumour to the culture to be assessed. The effect of addition of NGF is shown in Figure 3.3.

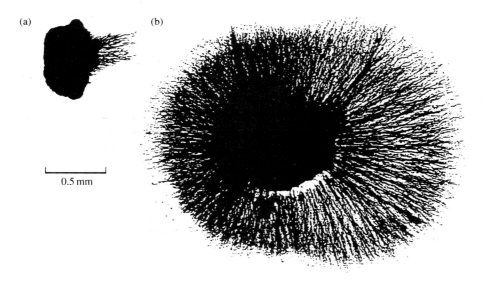

Figure 3.3 The appearance of a dorsal root ganglion from an E8 embryo in culture: (a) in the absence of NGF, and (b) in its presence.

At the start of the search for the identity of NGF, the effect of treating the tumour extract with enzymes was investigated to discover whether different components of the extract might be more active than others. Snake venom was used, since it contains very powerful enzymes that could possibly break down proteins in the tumour extract. To everyone's suprise the snake venom itself proved to be a much more potent growth-stimulating and cell-survival agent than the tumour cell extracts! This serendipitous result prompted Levi-Montalcini to search for a more abundant source of the agent. As snake venom is synthesised in the salivary glands, she looked at the salivary gland in a common laboratory animal, the mouse, as a possible source for NGF. The mouse salivary glands indeed provided a plentiful source of NGF for purification, although the role of NGF in the salivary glands remains a mystery.

NGF was purified, isolated and characterized as two identical chains of 118 amino acids. Further experiments demonstrated that NGF is released by relevant targets during development and that axons innervating those targets possess NGF receptors. Radiolabelling experiments have shown that NGF is taken up by the axon terminals and transported back to the cell body. Treatment of embryos with anti-NGF antibodies causes almost complete loss of sympathetic ganglia and loss of some neurons in the dorsal root ganglia. On the other hand, increasing the amount of NGF in the embryo spares some neurons that would otherwise die during development.

The action of NGF in promoting neuronal survival shows that it is a chemotrophic factor. The stimulation of neurite outgrowth by NGF in living organisms and in tissue culture might suggest that it could also guide axons by acting as the attractant in chemotropic guidance (see Section 2.4.5). However, in tissue culture it only affects outgrowth from ganglia that have already innervated their targets by the time of their removal from a chick between E6 and E8; it has no effect on neurite growth from younger ganglia. Also NGF is not present while sympathetic or dorsal root axons are growing, and neither type of axon has NGF receptors until they reach their targets in normal development. These observations suggest that NGF is involved in promoting cell survival and axon outgrowth only *after* the target has been contacted.

Since the pioneering studies on NGF, other chemotrophic substances have been isolated. Like NGF, they act only on particular types of neuron: they are only released by the targets of these neurons. Most seem to become important in cell survival once the target has been reached. These factors, like NGF, do not guide axons to their targets, but seem to play a role in matching the number of neurons to the size of their target.

3.2.3   Neuronal survival—making the right connections

As noted above, cell death is concurrent with synapse formation. It could therefore provide a mechanism for ensuring that only those neurons that innervate appropriate targets survive. There is good evidence for such selection in the mammalian visual system, where one of the consequences of cell death is the formation of a more accurate visual projection.

In normal mammals about 50% of retinal ganglion cells die after they have made contact with targets in the brain (between two days before birth and post-natal day

12 (P12) in the rat). As in the spinal cord, this is probably a mechanism for matching the number of retinal ganglion cells with the number of their target neurons in the brain. However, recent tracer studies have shown that selective cell death also may be involved in ensuring that only those cells that have reached their appropriate targets survive, whereas those that have reached inappropriate targets are eliminated.

Dennis O'Leary (1987) used a long-lasting retrograde tracing dye (Book 2, Section 8.7.1) to look at changes in the axonal projections from the retina to the caudal region of the superior colliculus during development of the rat. When he injected the dye into the caudal region of one superior colliculus at P12, and looked for dye-filled cells at P14, he found the pattern shown in Figure 3.4a. Most of the dye-filled cells are found, as expected, in the nasal region (N) of the contralateral retina. Figure 3.4b shows the distribution of cells in the retina when the injection was done at P0 (day of birth) and the retina examined two days later. There are many more filled cells scattered over the retina outside the nasal region. This distribution is more widespread than when injection took place at P12: there are retinal ganglion cells scattered throughout the retina that had contacted the caudal superior colliculus. What happens to these cells?

The advantage of using a long-lasting dye in these experiments is that animals could be injected at P0 and the distribution of dye-filled retinal ganglion cells could then be observed at P12, as shown in Figure 3.4c.

☐   What does the distribution of filled cells shown in Figure 3.4c suggest has happened?

■   The distribution looks like that seen in animals injected at P12 and observed at P14 (Figure 3.4a). The small number of filled cells outside the nasal region suggests that most of the retinal ganglion cells that had contacted the inappropriate region of the superior colliculus (Figure 3.4b) *must have died* between birth and P12.

Thus, the localized projection from the nasal region of the retina to the caudal region of the superior colliculus is due in part to the death of non-nasal retinal neurons which projected to the caudal colliculus.

Some 25–30% of retinal ganglion cells fail to reach their correct targets in the brain, and hence die. This leaves the death of a further 25% of retinal ganglion cells to account for in order to make up the known loss of 50% of retinal ganglion cells. These other neurons innervate their correct targets, so selection among them for those that should die is presumably based on a mechanism for matching the number of neurons to the target size, such as competition for a survival factor.

However, neural activity also plays a role in the control of cell death, though it is difficult to demonstrate this in the visual system. The neurotransmitters in the visual system have not been identified, and it is difficult to apply possible blocking agents reliably to the central nervous system for extended periods of time. Activity can be abolished by inhibiting action potentials, however. This is achieved by treating the retinal axons with tetrodotoxin (TTX), which selectively blocks the voltage-gated sodium ion channels in the axon (Book 2, Section 3.6).

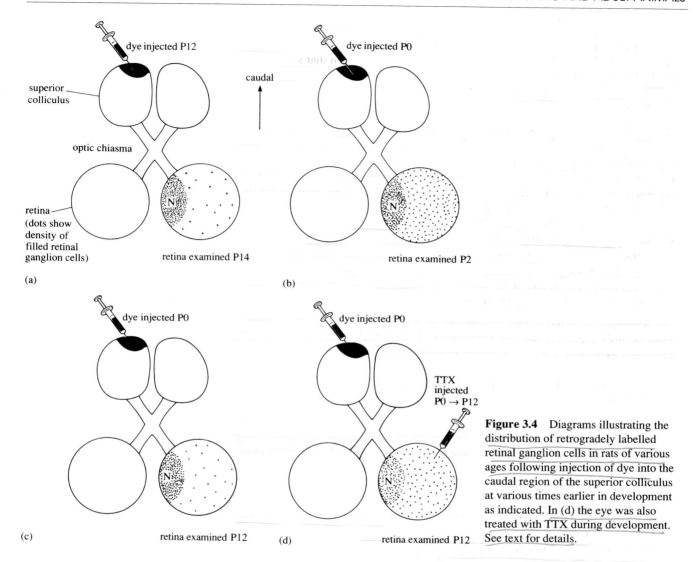

Figure 3.4 Diagrams illustrating the distribution of retrogradely labelled retinal ganglion cells in rats of various ages following injection of dye into the caudal region of the superior colliculus at various times earlier in development as indicated. In (d) the eye was also treated with TTX during development. See text for details.

If TTX is injected into one eye at regular intervals, it prevents activity in its optic nerve. Injection during the period of developmental cell death does not affect the overall loss of cells. However, treatment does affect the *pattern* of cell death. In animals with action potentials inhibited by TTX, cell death takes place evenly throughout the retina, and does not therefore result in a more accurate retinal projection, as would normally be the case (compare Figure 3.4c with Figure 3.4d). These results show that the activity of the retinal neurons must play a role in the identification of which cells are to die and which are to survive.

Although cell death in the visual system seems to play a role in the accuracy of the targeting of the retinal axons, this does not seem to be true for the motor projection to limb muscles. The motor projection can be analysed by injecting individual muscles with tracer and looking at the distribution of filled motor neurons in the spinal cord. Each muscle is innervated by a cluster of motor neurons which occupies a characteristic position within the ventral horn. The pattern of axons innervating various limb muscles looks very similar before and after the period of cell death. There are no motor neurons that have obviously reached inappropriate target muscles in the limb early in development which are removed later by cell death. So, here again, neuronal survival appears to depend on successful competition for a survival factor.

The apparent difference in the function of cell death in visual and motor systems serves to emphasise that different neuronal systems use different mechanisms to achieve accuracy of projection.

3.2.4 Neuronal survival—the influence of inputs

Muscle innervation and the visual projection show that the survival of neurons depends on functional synaptic contact with their targets. Neuronal survival also depends on *receiving* functional contacts from sense organs or other neurons. For example, the removal of peripheral sense organs causes the death of brain neurons to which the sense organs normally project. The removal of one eye in the early chick or amphibian embryo results in a loss of neurons in the contralateral optic tectum. Similarly, the removal of an eye in new-born mice is followed within a week by massive cell death in the contralateral superior colliculus, and in the cell layers in the lateral geniculate nucleus of the thalamus which would have received projections from the removed eye (Book 3, Section 4.6.1).

In these examples the neurons affected are those that are contacted *directly* by the sensory afferents.

A group of neurons in the mouse cerebral cortex is innervated by axons from the whiskers. These neurons are called the *whisker barrel field.* Neuronal survival in the whisker barrel field is affected by removal of sense organs (that is, whiskers) in the periphery.

☐ Do sensory axons in the periphery make direct (that is, monosynaptic) contact with cortical neurons?

■ No. There are at least two synapses between sensory afferents in the periphery and the cortex. One of these must be in the cranial nuclei for inputs from the head, or dorsal column nuclei for inputs from the body. Another of the synapses is in the thalamus. Neurons from the thalamus project in turn to the cortex (Book 2, Section 9.5.1).

☐ How is the input from the whiskers arranged in the cortex?

■ You should recall from Book 2, Section 8.8.2 that inputs from each whisker terminate in a special area within layer 4 of the somatosensory cortex called a barrel.

The arrangement of 'whisker' barrels corresponds to the arrangement of whiskers on the muzzle of the animal (this area is the maxillary region that you met in Section 2.4.5). If whiskers are removed at birth (before the whisker barrels have formed in the cortex), those barrels that correspond to the deleted whiskers will fail to develop (Figure 3.5).

This kind of interactive mechanism allows changes in the size of the peripheral organ to be accommodated by the nervous system. This is nicely illustrated by the

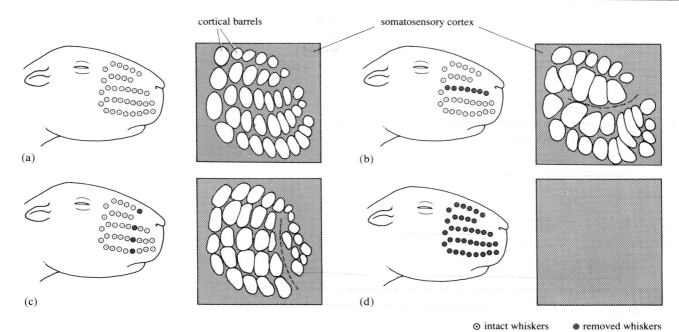

Figure 3.5 A schematic representation of the relationship between the whiskers on the muzzle of a young mouse and the arrangement of barrel fields in layer 4 of the somatosensory cortex. (a) shows an intact animal. In (b), (c) and (d) selected whiskers were removed at birth, leading to degeneration of the corresponding barrels (broken lines in (b) and (c); total disappearance of barrels in (d)).

observation that, in animals genetically endowed with extra whiskers, there is an extra barrel for each within the barrel field. Here the presence of the extra whisker in the periphery results in the survival and differentiation of additional neurons in the cortical barrel field.

☐   Bearing in mind that sensory axons do not make direct synaptic contact with cortical neurons, what is the significance of these whisker barrel data?

■   Whiskers are several synapses away from the whisker barrels in the cortex. Therefore, whatever the factor from the whisker that promotes survival of whisker barrels, it can exert its effects through several synapses.

Summary of Section 3.2

Neurons die if they do not make synapses with target cells. Among those neurons that do make contact with targets, there is competition for some kind of survival factor (chemotrophic factor) that is released by the target cells, picked up and then transported back to the cell body of the innervating neuron. If neurons receive insufficient survival factor, they die. The molecular identity of several likely chemotrophic factors has been discovered. The observation that cell death may also eliminate neurons that innervate inappropriate targets suggests that cell survival is a more complex affair than neurons simply doing battle with one another for a target-derived elixir of life! Once the period of neuron generation is over, the fine adjustment of neuronal number is brought about by selective cell survival. Both the afferent input to neurons and their target contacts influence the survival of neurons.

3.3 The survivors—differentiation and synaptogenesis

The preceding sections described how the number of neurons is regulated by both the target connections and input connections. The contacts that the surviving neurons make with their targets and the inputs they receive from other neurons also play a major role in the way they continue to differentiate, influencing the details of their axonal branching patterns and also their dendritic structure. These aspects of the further maturation of surviving neurons are addressed in this section.

3.3.1 The effect of afferent inputs on differentiation

There are many examples demonstrating the importance of normal inputs to neurons for their differentiation as well as their survival. In the whisker barrel field the presence of barrels depends on the presence of the whisker in the periphery. The structure of the whisker barrel is very characteristic and complex. Is this pattern the result of interaction between thalamic afferents and cortical neurons? Or is the pattern an intrinsic property of particular whisker barrel neurons in the cortex, which thalamic afferents then contact? The answer appears to be that thalamic afferents induce groups of cortical neurons to form barrels.

If the somatosensory area of the mouse cortex is replaced by cortical tissue from a different area (for example, the visual cortex) on the day of birth, before the thalamic afferents arrive in the cortex, the transplanted tissue clearly develops all the special structures characteristic of the barrel field. In the converse experiment in which tissue is moved from the barrel field area to another region of the cerebral cortex, barrel fields fail to develop in the transplanted tissue. In these experiments the organization of neurons and the pattern of their dendrites within different regions of the cerebral cortex seems to depend on the presence of particular thalamic afferents.

3.3.2 The influence of peripheral inputs on synaptogenesis

The stretch reflex is a good example of the precision of connections in the nervous system.

- ☐ What are the basic features of the neuronal circuit of the stretch reflex?

- ■ The sensory axons from a muscle stretch receptor (muscle spindle) make precise excitatory connections with the α motor neurons that innervate the same muscle. Stretching the muscle causes a reflex contraction of the same muscle (Book 2, Section 9.3.3).

Experiments conducted by Eric Frank and his colleagues at the University of Pennsylvania (Frank, Smith and Mendelson, 1988) in the developing frog show how contact with muscles in the limb defines the connections that dorsal root axons make with motor neurons in the spinal cord. The neurons of which the stretch reflex is composed can be revealed by exposing a muscle nerve to horseradish peroxidase (HRP). HRP will be taken up by the motor neurons to the muscles as well as the sensory axons that come from muscle spindles in the same muscle (Figure 3.6), since the axons of both are part of the same nerve. If other muscle nerves are filled, considerable overlap in their terminal arbours (that is, the tree-like pattern of all their branches) is apparent. Despite this overlap, there is great precision in the actual connections between sensory afferents from the muscle spindles and their motor neurons. This precision can be demonstrated by recording the synaptic potentials in motor neurons with intracellular electrodes.

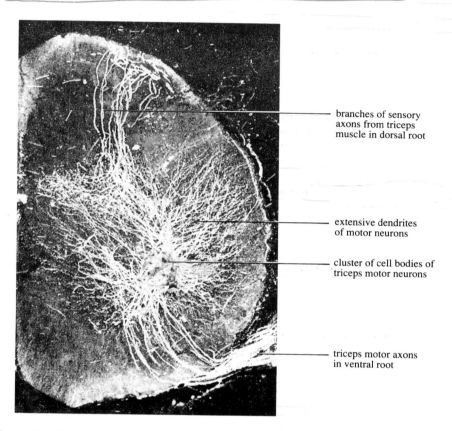

branches of sensory axons from triceps muscle in dorsal root

extensive dendrites of motor neurons

cluster of cell bodies of triceps motor neurons

triceps motor axons in ventral root

Figure 3.6   Cross-section of half the spinal cord of an adult frog after HRP take-up by the triceps nerve. Axons and cells containing HRP appear white. Sensory axons of the dorsal root are at the top of the figure; motor axons of the ventral roots are at the bottom of the figure. The central area contains the extensive motor neuron dendrites.

☐   What kind of potentials would you record from a motor neuron when (a) the muscle spindle sensory axons from the muscle innervated by the motor neuron were stimulated, and (b) the muscle spindle sensory axons from the antagonistic muscle were stimulated?

■   (a)   Stimulation of the muscle spindle sensory axons from the muscle spindles will produce *excitatory* postsynaptic potentials (EPSPs) in the motor neuron that innervates the same muscle.

(b)   Stimulation of the muscle spindle sensory axons from the antagonistic muscle will produce *inhibitory* postsynaptic potentials (IPSPs).

The appropriate EPSPs and IPSPs produced in motor neurons in response to stimulation of sensory axons from various muscles reflects the accuracy of these connections. The EPSPs produced by stimulation of sensory axons in the nerve supplying the muscle that the motor neuron innervates is always much larger than those produced when sensory axons in other muscle nerves in the limb are stimulated. Thus, sensory axons from sensory receptors in the triceps muscle almost exclusively excite motor neurons that innervate the triceps muscle and not those innervating other muscles such as those in the shoulder or chest.

This precise connection between the sensory axons and motor neurons might arise from particular sensory neurons synapsing only with particular motor neurons through a process of specific guidance. In order to test this hypothesis, sensory axons that normally innervate thoracic rather than limb muscle were made to innervate the forelimb of the frog. This was accomplished by transplanting the dorsal root ganglion from the thoracic region to the forelimb in the tadpole. The reflex connections were tested in the adult frog when the limb was fully grown.

Figure 3.7 shows the results: sensory axons from the thoracic dorsal root ganglion that innervated the triceps muscle made most excitatory connections with triceps motor neurons, as in normal animals. Stretching the triceps muscle excited afferents, which produced large EPSPs only in triceps motor neurons. In this situation, sensory axons from dorsal root ganglion cells that would not normally innervate a limb had made synaptic contact selectively with the appropriate motor neurons in the spinal cord.

This result shows that synapses made by sensory axons within the spinal cord are defined by the targets that they innervate in the periphery.

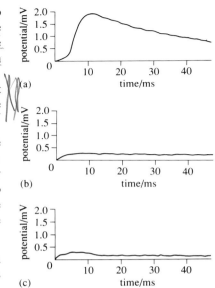

Figure 3.7 Synaptic responses evoked in triceps motor neurons by stimulation of sensory axons from the triceps muscle in a frog in which a dorsal root ganglion had been transplanted from the thoracic region when the frog was a tadpole. As in normal animals, stimulation of the sensory axons from the triceps muscle (a) results in a clear EPSP in the triceps motor neurons and not in motor neurons innervating other muscles, such as the shoulder (b) or chest muscles (c). The sensory axons were stimulated at time 0.

3.4   The refinement of axonal projections

Developing neuronal connections do not show the precision characteristic of the mature animal. Section 3.2 described how an overabundance of neurons is pruned down, enabling populations of interconnected neurons to be matched in size and type. This section considers how the precision of axonal connections is refined during later neuronal maturation without the death of neurons.

3.4.1   Selective pruning of axon branches in the cerebral cortex

As you saw in Section 3.2.3, selective cell death in the visual system is partly responsible for the precision of connections between the retina and the superior colliculus. Selective loss of some axon branches is also a factor in the achievement of appropriate connections between neurons in the developing cerebral cortex.

This phenomenon is nicely illustrated by the developing pyramidal tract in the rat.

☐   Where in the cerebral cortex would you expect to find cells whose axons project to the spinal cord along the pyramidal tract?

■   The pyramidal tract contains the axons primarily from layer 5 pyramidal cells in the motor cortex (Book 2, Section 8.8.5).

Two days after injection of a dye into the pyramidal tract at P20 it was found that layer 5 pyramidal cells in the rostral part of the rat cerebral cortex contained the dye. However, if the injection is made at P2, dye-filled cells are found at P6 in both the rostral cortex and the caudal cortex, including the visual cortex. When an animal is injected at P2 and examined at P25, the dye-filled cells are still found in both the rostral cortex and the caudal cortex, as they had been at P6. The observations are summarized in Table 3.1.

Table 3.1   Location of cortical cells containing dye following injection into the pyramidal tract.

Age when dye injected into pyramidal tract/days	Age when cortical cells observed for presence of dye/days	Location of cortical cells containing the dye
P20	P22	layer 5 pyramidal cells in the rostral cortex
P2	P6	layer 5 pyramidal cells throughout the cortex
P2	P25	layer 5 pyramidal cells throughout the cortex

☐   What do these results suggest?

■   The cells in the rostral and caudal parts of the cortex, including the visual cortex, send axons into the pyramidal tract early in development. Although these connections are lost later (there is no labelling of cells in the caudal cortex from injection at P20), the parent cells are still present at P25. Thus, in this example, *axon branches* are lost, not the whole neuron.

In further experiments, two different dyes were injected into the pyramidal tract: one dye was injected at P2 and the other at P25. When the visual cortex was examined, it was revealed that, at P2, cells in the visual cortex sent axons both to their normal destinations, such as the superior colliculus, and also down the pyramidal tract. By P25 the 'erroneous' projecting branch in the pyramidal tract had been lost. It seems likely that functional connections with both the appropriate and inappropriate targets compete, and that the appropriate connection 'wins'.

The survival of inappropriately projecting axon branches can be changed by various procedures that alter the normal innervation pattern. There are many extra axonal connections early in development between the two visual cortices; these redundant connections are later lost (Figure 3.8a and b). In the mature animal only the parts of the visual cortex containing cells whose receptive fields lie in the binocular field of view are connected through the corpus callosum (Book 2, Figure 8.10d). If one eye is removed at birth, many callosal axons that would normally be pruned are retained (Figure 3.8c). Thus, removal of afferents from one eye to the visual cortex results in the survival of a number of callosal afferents to the *same part of the cortex*. When the eye is present and there are lots of afferents to that part of the cortex, the callosal axons compete and many are lost. When the eye is absent, there are fewer inputs to that part of the cortex and many callosal axons survive. Here again the refinement of neuronal connections involves some kind of competition between axon terminals.

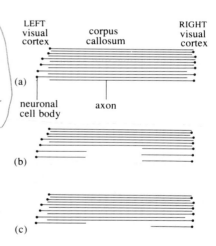

Figure 3.8   Diagrammatic representation of axon projections from left and right visual cortices, through the corpus callosum, to the equivalent place in the opposite visual cortex: (a) immature state—lots of projections in opposite directions; (b) mature state—only projections from areas of the visual cortex with binocular receptive fields remain; (c) mature state after removal of one eye early in development; there has been little pruning of inappropriate connections, and the immature state remains (note the similarity in number of projections to those in part a).

3.4.2 Fine tuning of axonal connections

Just as there is evidence that aberrant axonal branches are selectively removed, so also there is evidence that the arbours of axons within their targets become more compact. Initially, axons have extensive arbours in their targets, but these gradually contract. The contraction, or refinement, of arbours depends on the activity and connections of the individual branches, and is sometimes referred to as *functional tuning*.

Many mammals are born with their eyes closed and with their visual system still immature. In the next chapter, experiments that demonstrate how the pattern of connections in the visual cortex can be affected by visual experience are described. It seems to be a general phenomenon in the development of complex accurate neural connections such as are found in the visual system, that these are influenced by the pattern of activity occurring both pre- and post-natally.

☐ What pattern do the retinal afferents from the two eyes form in the lateral geniculate nucleus (LGN) of the thalamus?

■ In the LGN the optic afferents from each eye terminate in alternate cell layers, forming a pattern of stripes or bands (Book 3, Section 4.6.1).

The cell layers receiving axons from each eye emerge gradually as the terminal arbours of axons from one eye expand selectively within their targets, whereas side branches located outside the target are lost (Figure 3.9a). (Notice here that the target is very discrete: it is only a small part of the lateral geniculate nucleus of the thalamus.)

In many mammals, the alternating layers in the LGN are present at birth, so that it might seem unlikely that function plays a role in this segregation. However, retinal axons are spontaneously active in the dark and also in embryonic animals. Carla Shatz and Mike Stryker at the University of California investigated the structure of these layers in embryonic cats following nerve impulse inhibition of retinal cells, which was achieved by a continuous infusion of tetrodotoxin, TTX, into the eyes of the cats. In normal animals and control animals infused with saline (instead of TTX), the layers are clearly apparent at E56 (cats are born at about E65). In the TTX-treated animals, however, there were no layers at this time. The arbours from both eyes extended over the whole of the LGN (Figure 3.9b). Individual arbours were clearly much enlarged in the animals subjected to nerve impulse inhibition compared with the controls.

These results show that activity plays a role in refining the retinal terminal arbours in the lateral geniculate nucleus *even before birth*.

☐ What other interpretation for these results can you suggest?

■ The TTX treatment may have affected the development of the LGN. It could also affect the growth of axons by interfering with axonal transport, that is, the transport of substances within the axon.

There was no evidence that the treatment had affected the growth of the lateral geniculate nucleus. The size and structure of the LGN and other brain regions was normal in treated animals. Transport of substances along the axons also seemed to

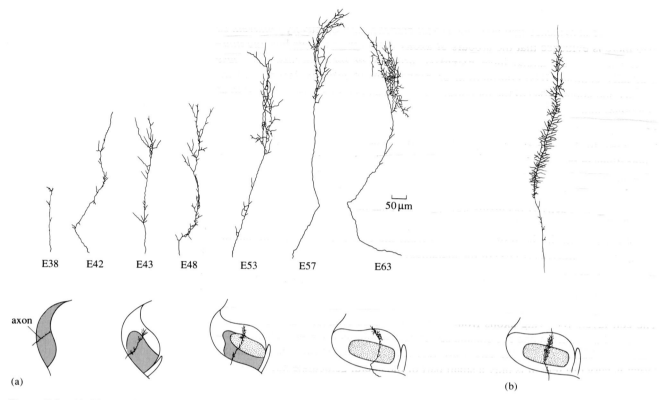

Figure 3.9   (a) Above: diagrams showing changes in the shape of the arbours of axons from the contralateral retina in the lateral geniculate nucleus in normal embryonic cats of different ages. Up to E53 small lateral branches are observable outside the target area, but by E57 they have been lost. Below: the outline of the lateral geniculate nucleus in transverse section, showing the position of the axons and their arbours. The shaded area shows input from both eyes. The unshaded area contains arbours from the contralateral eye, and the dotted area contains arbours from the ipsilateral eye. At E63 the segregation of the contralateral and ipsilateral arbours is complete. (b) The effect of TTX treatment on axonal growth; notice that there is no reduction in the axonal arbours.

be normal. The increase in the size of the axonal arbours in treated animals suggests that blocked axons could continue to grow. Shatz and Stryker were therefore justified in concluding that spontaneous action potential conduction was important in shaping the segregation of the retinal afferents.

Thus, activity in neurons, even in the absence of natural stimulation, influences the connections that a neuron makes with its target. The next section addresses the question of how activity can refine connections.

3.4.3   The mechanism of functional refinement of connections

In other visual projections (for example from the eye to the optic tectum in fish, and from the LGN to the visual cortex in mammals), the importance of neuronal activity in shaping the patterns of connectivity has also been demonstrated. If neuronal firing patterns are altered experimentally (for example by electrical stimulation, or by rearing the animal in a situation where the only light it sees is a brief bright flash once every 1–2 seconds), refinement of the connections does not occur.

Such experiments suggest that refinement must depend on functional patterns of neuronal activity (as opposed to simply random activity), and has led to the hypothesis of selective stabilization of synapses. This hypothesis proposes that synapses are strengthened when the presynaptic neuron and the postsynaptic neuron are active at the same time. Neurons are contacted by many synapses, but only those where the pre- and postsynaptic neurons are simultaneously active will be strengthened. This hypothesis of selective stabilization of synapses is remarkably similar to current ideas of what occurs during learning (see Section 5.4). The parallel was not lost on developmental neurobiologists. They began to test the effects of drugs known to block a particular kind of durable change in neuronal functioning, called long-term potentiation (see Section 5.7.3), on the refinement of axonal connections.

In both the optic tectum of a fish and the visual cortex in mammals, application of such blocking agents has a similar effect to inhibiting or changing neuron activity: the refinement does not occur. The mechanism of refinement of the visual projection may therefore be similar to that proposed to account for learning. In both phenomena, synaptic strength and stability is affected by the pattern of neural activity (Section 5.4).

What are the advantages of such a functional mechanism for the refinement of axonal projections? In fish and frogs the visual projection changes during development. In these animals, both the retinas and tecta increase in size throughout life as the animal changes shape and grows; as both young and old animals must possess effective visual systems, continual functional refinement is a necessity.

Functional refinement is also an ideal mechanism for the selection of precise neuronal connectivity in systems where there are inevitable differences between individuals due to environmental factors. Consider the input from a point within the binocular visual field (see Book 3, Section 4.6). A single cell in the visual cortex responds to a stimulus at one point in the visual field, yet the initial stimulus may be received by either eye. The part of each retina which responds to the point stimulus must project to the same single cell in the visual cortex, though the projection is not only to that cell. This represents an amazing feat of connectivity. The projection from each retina is accurate, despite inevitable variations between individuals in the sizes of the eyes, their position in the eye socket and the distance between the eyes, all of which are unpredictable and are produced by many factors, both genetic and environmental.

Summary of Sections 3.1–3.4

The first part of this chapter addressed a variety of issues surrounding neuronal survival and synaptogenesis. Broadly, these issues were:

1   the selective survival of neurons that innervated targets and successfully competed with other neurons for some survival factor produced by the target;

2   the influence of the target on synaptogenesis;

3   the elimination of functionally inappropriate synapses and projections.

The chapter now goes on to consider the extent to which the mechanisms occurring during development are present in the mature nervous system.

3.5   Growth of neurons in maturity

Is development finished when an animal is 'mature'? It is difficult to be precise about when development is complete because, as you have seen in the preceding sections, different parts of the nervous system have different timetables of development. There are a number of phenomena seen in mature animals (that is, when the animal is sexually mature) which strongly support the notion that development never stops completely. Most tissues retain some developmental capacity throughout the life of the animal. One indication of this continued capacity is seen in the response to injury. Many of the events that follow injury to the nervous system are reminiscent of those occurring during development. For example, axons re-grow and intact neuronal connections close to the site of injury reorganize themselves. Like the variation in the timetables of development between various parts of the nervous system and the variation between different animals, so too there are quite large variations in the response to injury.

Injury is rather a dramatic demonstration of the continued flexibility of neuronal connections, but the uninjured nervous system is also not a static structure. Tissue prepared for observation with a microscope provides a snapshot of connections which does not reveal the true dynamic nature of neuronal connections. Learning is one important sign of this continued flexibility.

Most of Section 3.5 is concerned with axon growth, whereas Section 3.6 examines the plasticity of synapses in the mature nervous system. Section 3.5 begins, though, with neuron birth in the mature nervous system.

3.5.1   New neurons for old?

During early development, removal of part of the embryo results in the replacement of the missing parts. Replacement usually involves neighbouring cells changing the way they differentiate and new cells being born. As development proceeds, so it becomes less likely that cells can alter their fate. To return to the rolling ball analogy presented in Book 1, Chapter 5, the further down the path of differentiation a cell is, the more unlikely it becomes that that cell will be able to switch to a new path.

A point of "no return"

The ability to replace missing parts differs widely, both among animals and also among different tissues within an animal. Many amphibians can regenerate complete limbs provided the limb is lost before metamorphosis (when the larval animal changes into a conspicuously different adult—for example tadpole to frog) is complete. Metamorphosis marks an important stage, after which regenerative ability is much reduced.

Some tissues in mammals are also able to regenerate (for example liver and muscle). You perhaps have personal knowledge of how muscle mass decreases with disuse and increases with use. The increase in muscle bulk is brought about by both an increase in the size of individual muscle cells and the formation of new muscle cells. The new muscle cells acquire motor innervation from sprouts from axon terminals at nearby neuromuscular junctions, not from new motor neurons. This shows that some neuronal connections can change in mature animals.

Size of muscle controlled by size of individual muscle cells + formation of new muscle cells.

☐   Can neurons divide in the mature nervous system?

■ No. As was discussed in Section 2.2.1, once neurons have gone through their final cell division, they cannot divide again.

If the central nervous system is damaged during the period when neurons are still being born, the missing cells are often replaced. After this time, however, cells lost through damage are not replaced. As different regions of the brain have different time periods of development, the time during which neurons cease dividing in the various regions will also differ.

There are some exceptions to the rule of lost neurons not being replaced. The lizard's tail is a good example. The end portion of the tail breaks off relatively easily, providing the animal with a means of escape from a predator that grabs it by the tail. Following the loss, a complete new tail, including muscles, vertebrae and spinal cord regenerates from the broken stump. The connections to and from the rest of the central nervous system are also re-established. Another example is the ability of the eyes of many fish and amphibians to regenerate from a small piece of the dark pigmented cells at the back of the eye. In many fish and amphibians, the eye, like the rest of the body, continues to grow throughout life. The eye gets bigger by the addition of new retinal cells around the eye margin and the axons from the added retinal ganglion cells grow to terminate in the brain. Here development continues until death.

In mammals the only example that is known where neurons continue to be added, and also to grow their axons to central targets throughout life is the olfactory system (Book 3, Chapter 1). The olfactory receptors are located in the nose in a delicate membrane that is being continuously destroyed. Thus, the olfactory receptors have to be replaced throughout life in all mammals. Their axons grow along the olfactory tract to their targets in the olfactory bulb at all ages.

The recovery of function following injury will therefore depend very much on the timing of the injury and also its location. In general, mature nervous systems do not replace lost neurons as part of their response to injury. However, there is often a large increase in glial cells at the site of injury, growing what is known as a glial scar.

The growth, or re-growth of damaged axons is considered next.

3.5.2 Re-growth of peripheral axons following injury

In mammals (including humans), if a peripheral nerve is crushed severely enough to sever the axons, the movements of muscles and sensations from the skin dependent on the nerve are lost. After some time, which depends on the distance between the damage and the targets innervated, the muscles will begin to move and sensation will eventually return. Though the crush severs the axons, it does not destroy the continuity of the glial cell sheaths; the severed axon leaves a hole through the glial cell sheath, which is effectively a hollow tube. After crushing (Figure 3.10b), many axons simply regrow down the glial cell tubes that lead to their original targets (Figure 3.10c).

The recovery when a peripheral nerve is actually cut (Figure 3.10d) rather than just crushed is nothing like as successful. There are thus limits to the ability of the peripheral nervous system to recover from injury. Microscopically, the axons can be seen to start to re-grow, but they often form dense tangles of axons which never reach their targets (Figure 3.10e). Following a cut there is no continuity of the glial cell tubes. Surgeons can improve the chances of successful axon re-growth by carefully joining the severed nerves together. Sometimes they insert an artificial bridge between the cut ends.

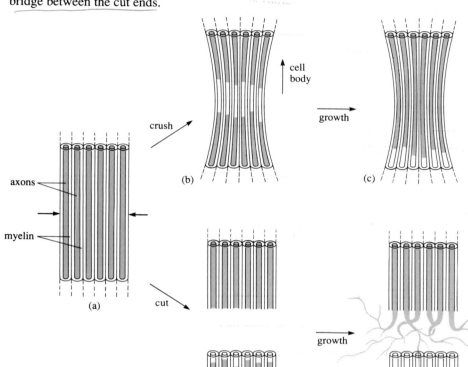

Figure 3.10   (a) Part of a nerve before crushing or cutting at the point indicated by the arrows. (b) Crushing severs the axons, but leaves the glial cell sheaths intact. (c) The same nerve after re-growth of axons into the glial cell sheaths; only proximal axons—that is, those attached to the cell body (indicated as at the top of the diagram)—re-grow; distal axons—that is, those on the side of the crush furthest from the cell body—degenerate. (d) Nerve soon after it has been cut; the nerve is under slight tension, so when it is cut, the cut ends spring apart, leaving a gap. (e) Proximal axons grow and sprout into the gap, where they become entangled in the glial scar; a few valiant axons may make contact with the distal glial cell sheaths, from which the distal axons have already disappeared.

☐   What would make a good pathway for axon growth across such a bridge?

■   As discussed in Section 2.4.2, axons grow well on a laminin substrate.

Nerve bridges used by surgeons are often made from tissues rich in laminin, such as the tissue surrounding muscle fibres (this laminin is present throughout life, and is probably related to the ability of muscles to increase in bulk and attract new motor axon branches locally). Plastic tubes containing glial cells have been used successfully too.

63

When such bridges are present, many more axons grow and innervate targets after cutting than do so in the absence of bridges, though these axons often fail to connect with their original targets because there are no specific guidance cues present. However, people (for example accident victims) whose nerves have been damaged can learn to recover function and interpret changed sensory information and motor connections if aided by careful physiotherapy, even though the precision of reconnection is poor.

3.5.3 Re-growth of axons in the central nervous system of amphibians and fish

If the optic nerve of a fish or an amphibian (such as a frog) is cut or crushed, the part of the optic nerve that is connected to the neuronal cell bodies in the eye (the proximal optic nerve) starts to grow: it regenerates. However, the part of the optic nerve that is not connected to the neuronal cell bodies in the eye (the distal optic nerve) degenerates and disappears. The proximal optic nerve regenerates well, makes connections in the tectum and eyesight is fully restored within 1–2 months.

☐ How is it that fish and amphibians are able to regenerate the optic nerve so effectively?

■ As was discussed above (Section 3.5.1), the retina continues to grow throughout life in fish and amphibians. The pathways and cues that guide retinal axons to their target regions in the brain must therefore be present throughout life, even though the surrounding areas of the brain are mature.

This capacity for accurate regeneration has been extensively used to investigate the cues guiding retinal axons to their tectal targets, beginning with the influential experiments of Roger Sperry in the 1940s. He was interested in how regenerating optic axons were able to find their targets in the tectum to give the animal what appeared to be perfect vision. A brief account of these experiments is given here.

Sperry tested the perfection of reconnection in frogs. Immediately after using delicate surgery to cut the optic nerve behind the eyeball, he carefully rotated the eye in its socket. He argued that if the connections reformed exactly as they had been before surgery, the animal would have an inverted view of the world. Behavioural tests carried out after the optic nerve had regenerated showed that this was the case: when a visual stimulus was presented to the operated eye, frogs always made systematic mistakes in their visually guided behaviour (Figure 3.11). The frogs never learnt to compensate for the visual misinformation. Histological observation of the cut showed a complex tangle of regenerated fibres, but despite this the axons from the eye had somehow re-grown and had made contact with their original targets on the tectum. These experiments led him to propose his *chemoaffinity* theory of neuronal specificity (Sperry, 1963). This theory sought to explain the precise matching of an axon with its target in terms of specific recognition between the axon and its target neurons. He suggested that a particular axon had a specific affinity for its target neurons so that it selectively formed synapses only with those neurons. Subsequent work has shown that connections made by growing axons change during development and after damage (see Section 3.6), so that the selective affinity of an axon for particular targets must change during development. Although these later observations rule out a rigidly specified system

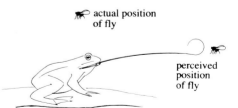

Figure 3.11 Misdirected strike at an overhead target by a frog after eye rotation. The insect is 'seen' in front when in reality it is located above the animal.

of axon–target connections, the idea that neurons establish their connections through active selection processes is still very much alive.

3.5.4 Re-growth of axons in the central nervous system in mammals: the role of glial cells

The success of regeneration in fish and amphibians contrasts sharply with what happens in mammals and birds. Cutting or crushing the optic nerve in a mature mammal results in a massive multiplication of glial cells around the damaged region, forming a dense glial scar. Although, as Cajal noted in 1928, a few axons seem to make 'small and frustrated generative acts', none grow through the scar.

There is evidence that axons can regenerate in immature brains, whereas in the mature animal they do not. In the 1980s an American scientist, Jerry Silver, set out to discover the reason for this difference. He first cut the corpus callosum (which connects the two cerebral hemispheres) in mice a few days after birth. Even when this operation is performed on P1 animals, the callosal axons do not manage to re-grow, but form a tangle of fibres either side of the cut (although they do regenerate if the operation is done before birth).

He therefore provided the axons with a bridge made from a special porous plastic. At various times after the operation he checked the re-growth of axons using tracer methods, and by looking at the appearance of the surface of the bridge using scanning electron microscopy, which provides a picture of the surface of the tissue. In animals operated on before P8 there was evidence that some callosal axons had regenerated across the bridge, establishing appropriate connections in the contralateral cortex. Operations performed at later times, however, resulted in dense tangles of fibres on either side but no growth across the bridge. Figure 3.12 shows a scanning electron micrograph of a bridge from an animal that had been operated on at P2 and examined 48 hours later. This shows the bundles of axons and clumps of rounded cells.

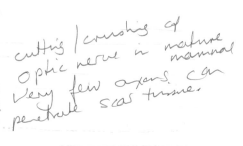

(handwritten note: cutting/crushing of optic nerve in mature mammal. Very few axons can penetrate scar tissue.)

Figure 3.12 Scanning electron micrograph of the surface of a plastic bridge inserted across the corpus callosum in a mouse at P2 and examined 48 hours later. The bundles of axons run from left to right, passing under a clump of rounded glial cells.

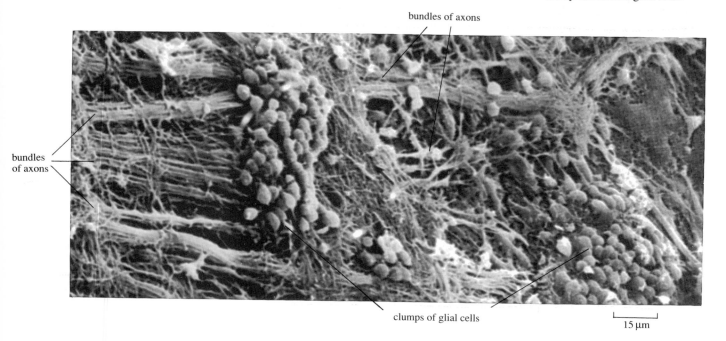

bundles of axons

bundles of axons

clumps of glial cells

15 μm

essay – regrowth of neurons!

Using immunohistochemical methods (Box 2.3), Silver showed that the rounded cells were glial cells. After inserting the bridge, the glial cells first migrate out of the cerebral cortex and colonize the bridge, followed soon after by the growth of axons. The glial cells in young brains are rounded and also have many extensions, some of which stick into the pores of the plastic bridge, whereas others extend among the growing axons. Bridges put into older animals (after P8) are also populated by glial cells, though these have a very flat rather than a rounded appearance.

The glial cells that invade bridges in young animals are obviously different from those in older animals. In a second set of experiments, Silver investigated whether glial cells from younger animals would promote axon growth in older animals. Bridges were removed from P2–4 animals and inserted into P17 or P34 animals after cutting the corpus callosum; these bridges contained numerous glial cells. Axon growth was noticeably more extensive under these conditions than in P17 or P34 animals with implants that did not contain glial cells. There was also a marked reduction in the glial scarring at the insertion point.

These results show two things—firstly, the importance of glial cells in axonal growth and, secondly, that glial cells, like all other cells in the body, change their properties as they differentiate. The improved growth of axons in older animals in the presence of young glial cells also implies that central nervous system axons can regenerate if provided with the right glial cell environment.

A further dramatic example of the importance of glial cells in the success of re-growth of central axons is seen in experiments in which central nervous system axons are provided with bridges made from peripheral nerve grafts. In 1981 Albert Aguayo cut the optic nerve in a rat just behind the eyeball, and carefully attached to the proximal end of the nerve a length of peripheral (sciatic) nerve taken from the thigh region of the same animal (Figure 3.13).

The result of tracer experiments showed two things. Firstly, in the presence of the graft, 10% of the retinal ganglion cells survived, compared with less than 1% survival in the absence of the graft. Secondly, a small proportion of retinal ganglion cells re-grew axons some distance along the peripheral nerve graft. Although the proportion of cells that grew into the graft was small (4%), it still represents a significant number, since there are about two million ganglion cells in the normal rat retina.

Are peripheral nerve grafts going to provide a wonderful new technique for the repair of injury to the central nervous system? Although this research has opened up a way to improve axon regeneration, there is still a long way to go before it can be hailed as a miracle method. Remember that only 4% of the retinal ganglion cells actually grew axons into the graft. At present there is some evidence that retinal axons, after growing along peripheral nerve bridges, can establish synaptic connections in a denervated superior colliculus. However, at the time of writing (1992), there is no evidence that such connections are either appropriate or that they can convey useful information to the animal.

The success of the peripheral nerve graft experiments prompted a rush to discover the molecular nature of factors that might affect the re-growth of axons. The interactions between retinal axon growth cones and glial cells from peripheral or central nervous tissue were observed in tissue culture. These observations revealed that growing neurites avoided contact with the glial cells that form the myelin

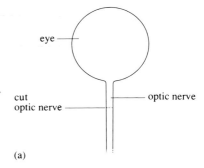

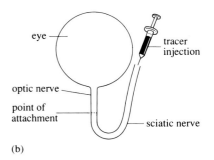

Figure 3.13 Illustration of the cutting of the optic nerve in a rat and the grafting of a length of sciatic nerve to the proximal end of the optic nerve. The site of injection of tracers to label cells in the retina retrogradely is also shown. (a) shows the eye before surgery; (b) shows the eye after surgery.

around axons in the central nervous system—the oligodendrocytes (Book 2, Section 3.8). The oligodendrocyte membrane contains substances that inhibit axon growth in tissue culture. The search for the identity of these substances is intense because of the possibility that antibodies to them might improve re-growth of central axons.

Another approach to discovering what factors might aid axon regeneration is to ask why regeneration is so successful in fish. Are the oligodendrocytes in fish different or does the difference lie in the axons? This question was answered by observing the growth of fish retinal axons on mammalian oligodendrocytes in tissue culture. The result is clear: the growth cones from fish retinal axons also avoid contact with mammalian oligodendrocytes. As in the peripheral nervous system, it is the difference in the properties of the glial cells rather than the difference in the properties of the axons which determines axonal growth. This conclusion is supported by the converse experiment: mammalian retinal axons grow well over fish oligodendrocytes. The differences between fish and mammalian central nervous system glial cells which have a bearing on axonal growth are not yet known.

These results emphasise that the growth of axons in vertebrates depends on the *kinds* of glial cells they contact. It is important, however, to realize that the success of fish optic nerve regeneration does not depend only on the presence of 'axon-friendly' glial cells. The success of fish optic nerve regeneration is due to three other important factors, all of which are directly related to the continued growth of the eye throughout life (Section 3.5.1):

1   Almost all the retinal ganglion cells survive for long periods of time after axon severance.

2   All surviving cells are capable of re-growing their axons.

3   Guidance cues are present so that axons can re-establish contact with their appropriate target cells in the brain.

This is a very different situation from the less than 1% retinal ganglion cell survival figure for mammals. Perhaps retinal ganglion cells are much more dependent on their targets in mammals than in fish.

3.5.5   Summary of axon regeneration

Just as for initial axon growth, the environment is all-important to the success of axon regeneration. Because of the obvious medical importance, a great deal of current research is directed towards improving neuron survival and axon re-growth. So far it is clear that, although axon growth can be improved in various ways, recovery of function will be limited unless guidance cues are present that will lead the regenerating axons to their correct destinations. From what little is known about axon guidance in normal development, it appears that timing and cell interaction play vitally important roles. Until the 'developmental clock' can be turned back ('the ball rolled up the hill again'), the success of nerve regeneration will be poor. Careful surgery, however, can improve the chances of some functional restoration after peripheral nerve damage.

3.6 Plasticity of synapses in the mature nervous system

Neuronal connections change during development, both in the normal course of maturation and also after various kinds of damage or interference. Is there any evidence for this kind of neuronal flexibility in mature animals, or are the connections in mature animals fixed and immutable? In Section 3.5.1, the ability of motor axons to sprout and innervate new muscle fibres produced by increased muscle use in mature animals was mentioned. Does such sprouting also occur in response to damage to muscles or axons? It is difficult to imagine that the dynamic interactions occurring between axons and targets during development are completely shut off in the mature animal. This Section examines whether such dynamic interactions do occur in maturity.

3.6.1 Synaptogenesis in the mature peripheral nervous system

If a peripheral nerve is cut or crushed in a mature animal, it will re-grow.

☐ What specific condition will assist regeneration of the peripheral nerve?

■ Contact with its appropriate glial cell sheath (Section 3.5.2).

While the damaged peripheral nerve re-grows towards its target, sprouting occurs in those intact motor and sensory axons that innervate the target close to the region that the damaged nerve had innervated. In this way the intact axons come to increase the area of the target that they innervate (Figure 3.14). What is the signal for this **collateral sprouting**? Various experiments suggest that the signal for sprouting is a chemotrophic factor released by the muscle. Under normal conditions the intact axons take up the factor, and transport it to their cell bodies, where it is broken down. When some of the axons have been damaged, more of the factor is present in the extracellular matrix, and this encourages nearby intact motor (and sensory) axons to sprout to innervate the muscle and its sense organs.

The loss of peripheral contact with targets can also affect the afferents to neurons. After cutting the axons of some motor neurons in the periphery, those motor neurons lose some of their afferent inputs, and other inputs become less effective at activating those motor neurons. These afferents are regained when contact with muscle (even if not the appropriate one) is re-established. It is possible that the chemotrophic factor discussed above might also play a role in determining the synaptic inputs to a neuron.

3.6.2 Silent synapses

Collateral sprouting has also been demonstrated in the central nervous system of fish, amphibians and mammals. Cutting one dorsal root (containing the sensory axons from a dorsal root ganglion to the spinal cord), for example, results in the axons from neighbouring dorsal roots sprouting to innervate denervated central neurons in the dorsal horn.

☐ How would you verify that sprouting had occurred?

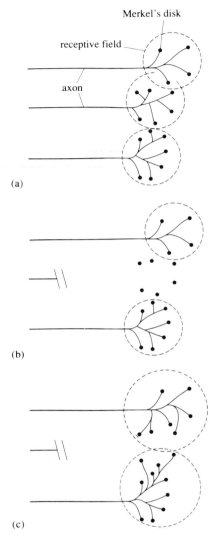

Figure 3.14 Collateral sprouting: (a) three sensory axons, each of which innervates a group of Merkel's discs; the broken circles represent the receptive fields of each of the neurons; (b) the middle axon has been cut and the distal part degenerates; (c) the axons above and below the cut axon have produced new collateral sprouts, which innervate the Merkel's discs originally innervated by the cut axon; the innervating axons have therefore increased the size of their receptive fields.

■ Sensory axons from the cut dorsal roots can be filled with dye. The extent of their terminations in the spinal cord can be compared with the intact situation, for example by looking at the equivalent dorsal root on the opposite side of the body. The pattern of connections of sensory inputs to the dorsal horn of the spinal cord can also be mapped using electrophysiological methods (Book 2, Box 8.3).

The sensory afferents form a topographically organized map in the spinal cord, the dorsal column nuclei, the ventrobasal nucleus and the somatosensory cortex (Book 2, Section 9.5.1). Pat Wall in London used electrophysiological mapping techniques to follow changes in the projections to the spinal cord of cats when dorsal roots were cut. Figure 3.15 shows the position of receptive fields recorded from dorsal horn neurons in a cat ; the dorsal roots receiving sensory input from one hind limb had been cut. On the intact side, as the electrode passes down through the grey matter, the receptive fields of cells encountered are in similar positions on the body; the thigh in this case. On the operated side a different picture emerges. Here the receptive fields of cells encountered in a similar part of the grey matter are very widely distributed.

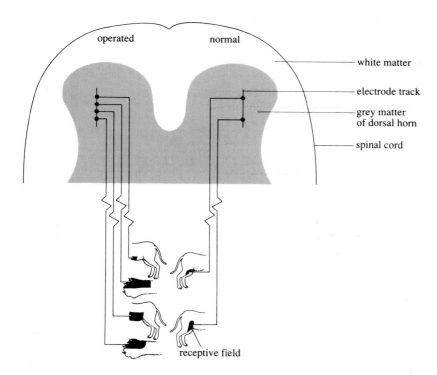

Figure 3.15   Schematic representation of the receptive fields of single dorsal horn neurons recorded at different depths along electrode penetrations through the dorsal horn of a mature cat, in which the dorsal root receiving input from one hind limb had been cut. On the right the penetration corresponds to the intact (right) side of the cat; on the left the penetration corresponds to the operated (left) side of the cat. Note that the recording positions in the operated side are closer together than those in the intact side of the cat, yet their receptive fields are further apart.

☐   How can you explain the changes that Wall found after cutting the dorsal roots?

■   The denervated dorsal horn neurons had acquired synaptic input from nearby intact sensory axons.

What was really surprising in these experiments was that the changes in the receptive fields were evident within hours rather than days or weeks after cutting the dorsal root. This was much too fast to be accounted for by a sprouting growth process (compare with the rate of growth seen in growth cones; Section 2.4.1).

These results were quite unexpected. Here was evidence that the 'wrong' connections must have been present all the time! The results suggest that in the normal mature animal the functional specificity of some sensory connections is a result of continuous inhibition of sensory information arriving via 'inappropriate' synapses. Cutting a dorsal root removes the inhibition so that previously inhibited (or 'silenced') synapses are now able to activate postsynaptic neurons.

So far, there is no good anatomical evidence for the existence of such silent synapses. It is very difficult to demonstrate that such connections are present; to do so, it would be necessary to record from *and* dye-fill *all* the contacts made between an identified axon and its target neurons.

3.6.3   Changing receptive fields of cortical neurons

Just as changes in the whisker innervation pattern of mice produced changes in the cortex several synapses away in immature mice (Section 3.2.4), so changes in peripheral sensory innervation patterns also produce changes (albeit limited) in the somatosensory projection in the mature cerebral cortex.

This has been demonstrated in the squirrel monkey (*Saimiri sciureus*) by Michael Merzenich, John Kaas and their colleagues, working variously in the University of California and Vanderbilt University, Tennessee. In anaesthetized animals, they used recording electrodes in the brain to make a detailed map of the receptive field area of a group of neurons in the somatosensory cortex. The receptive fields were on the ventral, or palmar surface of the hand (Figure 3.16a). They noted the position of the recording electrodes with respect to the pattern of blood vessels overlying the cortex. The median nerve was cut, 2–4 cm above the wrist (the median nerve innervates the ventral surface of the hand), and the proximal end was tied to prevent any proximal regeneration. They followed what happened to the receptive fields of the same cortical neurons by carefully remapping their receptive fields, using the pattern of blood vessels to position the recording electrodes. Figure 3.16a shows the areas on the hand that represent the receptive fields of the small group of neurons in the cortex.

Within 14 h of the median nerve being cut, the cortical neurons no longer responded to stimulation of what had been their receptive field area. Instead, there were new, much smaller receptive fields on both the dorsal and ventral surfaces of the hand (Figure 3.16b). Thus, the cortical neurons that had previously only responded to stimulation of part of the ventral surface of the hand (via the median nerve) now received some input from other areas of the hand, via the intact ulnar and radial nerves.

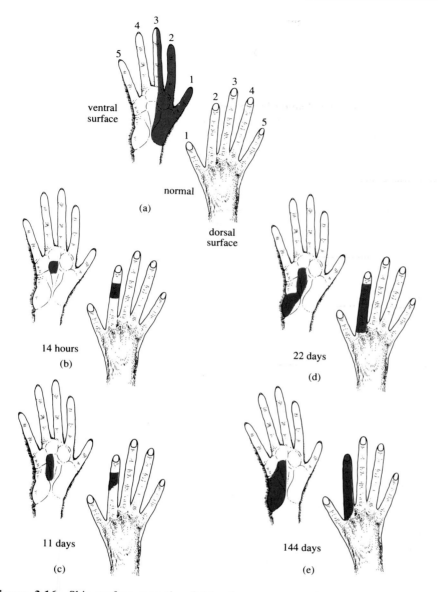

Figure 3.16   Skin surface receptive fields of some neurons in a small part of the somatosensory cortex of the squirrel monkey: (a) the normal receptive field area is shown shaded. The receptive field areas seen (b) 14 hours, (c) 11 days, (d) 22 days and (e) 144 days after the median nerve was cut are also shown.

☐   How might such an immediate change in the receptive fields of the cortical neurons be explained?

■   The most likely explanation is the unmasking of 'silent' synapses. In other words, the pathways were already present, but were inhibited by the input from the median nerve. When the median nerve was cut, the inhibition ceased and the new pathways were activated.

Over the next few weeks the new receptive fields increased in size (Figure 3.16c–e). By 144 days the area of hand projecting to the original cortical sensory neurons was quite extensive. What is interesting is that the changes in the receptive fields were not haphazard: they were still topographically arranged. This longer-term change is thought to be due to a modification of the thalamic afferents as they sprout to innervate—or in other ways start to excite—cortical neurons that previously responded only to median nerve inputs.

If the median nerve is not tied, it can regenerate and re-innervate the skin within a year. As this happens, the receptive fields of the cortical cells change again, and the new thalamic inputs are lost in favour of those carrying information from the regenerated median nerve. However, the detail of the pattern of connections never regains the original topography seen before the median nerve was cut, so the receptive field never quite returns to that mapped before the operation.

These observations show that even in the adult animal the nervous system is not a 'hard-wired', inflexible system. The impression of the nervous system as a complex, but static structure, which is gained from looking at fixed and sectioned specimens through the microscope is not at all like reality.

3.6.4   Observing structural changes in the living synapse

Routine histological methods of preparing tissue for microscopic examination leave the tissue dead and static. Yet the tips of growing axons are dynamic, both in living animals and in tissue culture. In tissue culture the internal contents of all cells are never still, and most cells extend and retract processes, and move. Until recently there was no method available to allow this movement to be observed microscopically in living animals. However, just such observations have now been made on the muscle end plates. Muscle end plates in some muscles can be easily and repeatedly exposed in anaesthetized animals. The end plates can be made visible by treating them with a mitochondria-specific fluorescent dye. This dye is non-toxic, and remains visible over several weeks. Figure 3.17 shows the appearance of a rat muscle motor end plate in photographs taken 4 months apart. Although the basic structure is similar, there are changes in the length of parts of the structure. Studies of end plates in ageing animals using normal methods of microscopy show that they become more complex with time, but this is the first clear demonstration of visible changes within a *single* identified end plate.

Does this image of restless synapses also apply to contacts between neurons in mature animals? Dale Purves in North Carolina has used non-toxic dyes to stain single cells in the superior cervical ganglion of the sympathetic system in the adult rat. The superior cervical ganglion in the neck can be exposed quite easily in an anaesthetized animal, and manoeuvred beneath a fluorescence microscope. Micro-pipettes containing the dye are then inserted into individual cells, and the dye is injected into the cell, filling its dendrites. After recording the shape of the dendrites using video, the wound is closed. Several days or weeks later the ganglion is re-exposed, and the cell, which still contains some dye, is re-injected and re-recorded on video. Figure 3.18 shows that after four days there are minor changes in the shape of the dendrites, but that after one month the changes are much more obvious.

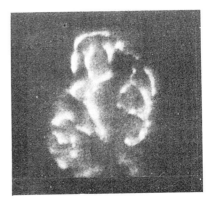
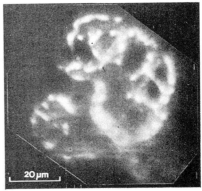

Figure 3.17 Photographs of a motor end plate in a rat muscle stained with mitochondrial dye. The lower photograph was taken 4 months after the upper photograph.

Although neurons of the sympathetic ganglia lie outside the central nervous system and have a different origin from cells within the central nervous system, it is very likely that all neurons are similarly dynamic. As various advances in techniques for observing neurons in living tissue become available, it should be possible to monitor the dynamic aspects of neuroanatomy which reflect ongoing changes in the structure and function of neurons.

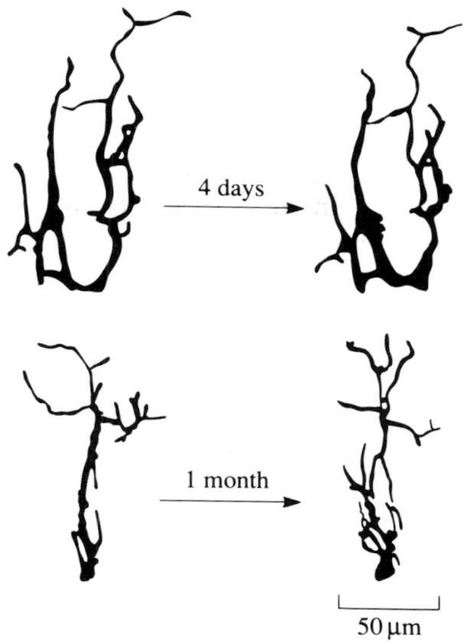

Summary of Sections 3.5 and 3.6

Evidence from intact animals suggests that connections remain plastic in mature animals. During axon re-growth, both central and peripheral axons produce growth cones whose behaviour is very similar to that seen during development. In some situations the glial cell environment permits growth but not guidance; in others both growth and guidance are frustrated by the glial cells.

The changes seen in intact parts of the nervous system following injury also demonstrate that the kinds of interactions between neurons that occur during development can be repeated, to a limited extent, in the mature nervous system. The effects on neuronal projections of cutting dorsal root axons makes the important point that changes in connectivity need not be associated with gross structural change.

It is perhaps not suprising that the mature nervous system can adapt to injury, since one of the fundamental properties of nervous systems is their responsiveness to changing inputs.

Figure 3.18   Temporal changes in the structure of dendrites of neurons in the superior cervical ganglion of the rat. The upper and lower pictures show two neurons left for different periods of time. The longer the interval, the more extensive are the changes in the dendritic pattern.

Summary of Chapter 3

This chapter has emphasized the interactions between neurons occurring as a result of their synaptic connections and activity, which influence their survival and growth. During this phase of neuronal maturation the precision of connections seen in the mature animal develops.

The first stage of the process is the matching of the number of neurons to the size of their targets through selective cell survival. In most parts of the vertebrate central nervous system there is an overproduction of neurons, most of which make contact with target cells. The selective survival of only a proportion of these neurons results in a downward adjustment of the number of neurons to suit the target size. In some systems, neuronal survival also plays a role in determining the accuracy of neuronal projections.

The continued refinement of neural connections involves changes in arbours such as the loss of aberrant branches and a change in terminal branching patterns. During this phase, neuronal activity plays a pivotal role. This is when alterations in the pattern of action potentials in neurons brought about by external events can alter the connections these neurons make.

During development, connections between neurons can change; they are particularly flexible or plastic at this time. Demonstration of this flexibility usually involves interference with normal development (for example the removal of some inputs to neurons where other cells or axons can take over the role of the missing inputs). In most cases this flexibility reflects the importance of neuronal interaction in shaping connections in the nervous system. Different neuronal connections exhibit very different timetables of development and extents of flexibility.

The flexibility of neuronal connections is not completely lost in the mature animal. Both the response to injury and the changes that occur during learning are reflections of the fundamental property of nervous systems that they can adapt to changing circumstances. It comes as no surprise to the developmental neuro-biologist that similar cellular mechanisms are involved in both the selective refine-ment of neuronal connections and also in the changes in neuronal connectivity that accompany learning, which will be described in the last chapter of this book.

The next chapter describes the effects that the processes discussed in this chapter have on the behaviour of the animal.

Objectives for Chapter 3

When you have completed this chapter, you should be able to:

3.1 Define and use, or recognize definitions and applications of, each of the terms printed in **bold** in the text.

3.2 Explain how both the contacts that a neuron makes with its target and the inputs it receives from other neurons can affect its survival. (*Questions 3.1 and 3.2*)

3.3 Explain how neuron survival can affect the precision of neuronal projections. (*Question 3.3*)

3.4 Describe the role of chemotrophic factors in selective cell survival and how such factors might be identified. (*Question 3.4*)

3.5 Explain the importance of neuronal contacts to the developing neuron. (*Questions 3.5 and 3.10*)

3.6 Explain the role of functional activity in refining the precision of neuronal connections. (*Question 3.6*)

3.7 Give an account of the regenerative capacity of peripheral and central axons in different types of vertebrate. (*Question 3.7*)

3.8 Explain the importance of the glial environment in axon regeneration. (*Question 3.8*)

3.9 Present evidence that the nervous system retains plasticity in the mature animal. (*Questions 3.9 and 3.10*)

Questions for Chapter 3

Question 3.1 (*Objective 3.2*)
Briefly describe how selective cell survival mediated by chemotrophic factors can help to set up precise and detailed correspondence between the numbers of neurons and the size of their targets in different parts of the nervous system.

Question 3.2 (*Objective 3.3*)
Although transplanting additional muscle tissue results in an increase in the number of motor neurons surviving in the nearby spinal cord, it never results in the survival of all motor neurons. Suggest possible reasons for this.

Question 3.3 (*Objective 3.4*)
What are the important criteria that must be met for a substance to be classified as a chemotrophic factor?

Question 3.4 (*Objective 3.5*)
In the experiments on the development of connections between dorsal root ganglia and motor neurons of frogs described in Section 3.3.2, it was very important to perform the operations well before metamorphosis in order to obtain normal reflex connections. What does this suggest?

Question 3.5 (*Objective 3.6*)
What is the significance of functional tuning in the development of the visual projection?

Question 3.6 (*Objective 3.6*)
Why do the visual pathways of fish have such a remarkable capacity for successful neuronal regeneration?

Question 3.7 (*Objective 3.7*)
Why do peripheral rather than central nerve grafts promote regeneration of axons? Do such grafts guarantee successful re-innervation of targets?

Question 3.8 (*Objective 3.8*)
Which mammalian glial cells facilitate axon growth?

Question 3.9 (*Objective 3.9*)
Until relatively recently the mature nervous system was considered to be a highly organized structure, which did not change. Why do you think such a view arose?

Question 3.10 (*Objectives 3.5 and 3.9*)
Does the refinement of neuronal connections necessarily involve the loss of all those that are inappropriate?

References

Frank, E., Smith, C. and Mendelson, B. (1988). Strategies for selective synapse formation between muscle sensory and motor neurons in spinal cord. In S. S. Easter, K. Barald and B. M. Carlson (eds), *From Message to Mind*, Sinauer Associates, pp. 180–202.

Levi-Montalcini, R. (1975). NGF an uncharted route. In F. G. Worden *et al.* (eds), *Paths of Discovery*, MIT Press, pp. 245–265.

O'Leary, D. D. M. (1987) Remodelling of early axonal projections through selective elimination of neurons and long axon collaterals. In G. Bock and M. O'Connor (eds), *Selective Neuronal Death*, Ciba Foundation Symposium, Wiley, pp. 113–130.

Sperry, R. W. (1963) Chemoaffinity in the orderly growth of nerve fiber patterns and connections, *Proceedings of the National Academy of Science, USA*, **50**, 703–710.

Further reading

Alberts, B., Bray, D., Lewis, J., Raff, M., Roberts, K. and Watson, J. D. (1994). *Molecular Biology of the Cell*, 3rd edn, Chapter 21: Cellular Mechanisms of Development (section on neural development), pp.1119–1137, Garland Publishing.

Brown, M. C., Hopkins, W. G. and Keynes, R. (1991). *Essentials of Neural Development*, Parts 4 and 5, Cambridge University Press.

Parnavelas, J. G., Stern, C. D. and Stirling, R. V. (eds) (1988). *The Making of the Nervous System*, Part 3: Molecules and guidance pathways, Part 4: Sharpening the pattern, Oxford University Press.

Purves, D. and Lichtman, J. W. (1985). *Principles of Neural Development*, Sinauer Associates.

CHAPTER 4
DEVELOPMENT OF BEHAVIOUR

4.1 Introduction

This chapter is concerned with those factors that, if present at some time during the development of an organism, affect the *later* behaviour of that organism. It is not specifically concerned with the ways in which the behaviour of one organism or one species changes with age, although Section 4.5 does address this point to a limited extent for humans. Rather, it is concerned with selected aspects of behaviour, and the factors during development which might influence them. Concentrating on the development of one aspect of behaviour at a time inevitably carries with it the risk of thinking that each different aspect of behaviour depends on a distinct developmental mechanism, isolated from the mechanisms that underlie other types of behaviour. It is therefore essential to remember that development involves a complex nexus of causal relations, with continuous interplay between organism and environment at every stage.

The development of behaviour is inextricably bound up with the development of the nervous system. As the nervous system grows and makes connections, so the possible behaviour patterns of the animal change. It follows from this that any factor that influences or alters the development of the nervous system may also alter behaviour. As you have seen in earlier chapters, the nervous system is undergoing considerable and rapid change in the early life of the organism, a feature that has two consequences, which together are sometimes referred to as *susceptibility* or *vulnerability*. Firstly, alterations to the development of the nervous system at one point in time can have a profound effect on the subsequent development of the nervous system, and hence behaviour. Secondly, any changes to the developmental path cannot usually be reversed in later life. These two consequences highlight the importance of Tinbergen's second way of answering the question, 'Why does an animal behave the way it does?', in terms of the development of behaviour. (Book 1, Section 1.1.1) A more precise question is: What factors during development influenced the eventual performance of the behaviour?

The development of behaviour in any given animal is dependent on a whole catalogue of factors, including for instance its genotype, its nutrition, the sensory stimuli with which it comes into contact and the hormones that are present. Only some of these factors have been studied, and then only in particular organisms. It is assumed that factors shown to affect development in one organism are general, and influence development to a greater or lesser extent in other species. The range of factors that can influence the process of development, and hence behaviour, is vast (this chapter presents only a sample of the possible factors), and the effect of those factors varies with their timing, both when they operate and for how long, and the species on which the factors are acting. Broadly, three classes of factor are considered in this chapter: (a) genotype, (b) external stimuli such as light and sound, and (c) hormones; these comprise the three main sections of the chapter.

The chapter concludes with a brief look at the extent to which these same factors influence human development.

The chapter begins by considering the importance of the genotype to the overall behaviour of the organism, a topic first introduced in Chapter 3 of Book 1.

4.2 Of genes and neurons

In Book 1, Chapter 3, a number of differences in behaviour between animals were attributed to differences in their genotype. For example, the differences in fighting behaviour shown by four strains of mice were related to their genetic differences, and the differences in speed of achieving mating by male yellow *Drosophila melanogaster* and male wild-type *Drosophila melanogaster* were also ascribed to genetic differences. The problem of how differences at the genetic level could be translated into differences at the behavioural level was briefly addressed only with respect to the testicular feminization (Tfm) mutation in mice. In that example, the receptor for the hormone testosterone in Tfm mice was seen to be inefficient at detecting testosterone, and so a range of biochemical activities that required testosterone as a trigger never occurred. The following section presents some examples where the effect of the genotype at the neuronal and behavioural level has been studied.

4.2.1 Gynandromorphs and courtship in *Drosophila*

During courtship in the fruit fly *Drosophila melanogaster*, males follow females around while beating their wings at the particular frequency that the females find attractive, and then, if the female is receptive, they will attempt copulation. Females do not show these behaviour patterns. One question that can be asked about these behavioural differences is: are there corresponding differences in the nervous systems of male and female *Drosophila*? A first step to answering this question was taken by Yoshiki Hotta and Seymour Benzer (1976), using the technique of genetic mosaics, which was briefly introduced in Book 1, Section 3.6.1.

Animals that have some cells of one genotype and other cells of a second genotype are called mosaics. A mature mosaic animal has patches of cells of one genotype interspersed with patches of cells of the other genotype. It also has some organs comprising cells of one genotype and other organs of the other genotype. The distribution of cells of the two genotypes throughout the animal is unpredictable. If one of the genotypes is female and the other genotype is male, then the resulting animal is a **gynandromorph**—part female, part male. Provided the two cell types can be distinguished in some way, then it is possible to establish which parts of *Drosophila* need to be male for the two behaviour patterns—following or wing vibration—to occur. (This technique only works in certain animals, such as insects, where individual cells express characteristics of their genotypic sex. In vertebrates, this does not happen; see Section 4.4.) The two cell types can be distinguished by **genetic markers**, proteins that one genotype, and hence one cell type, produces and the other does not; that is, the cells are phenotypically different. The marker might render the cell a different colour, or cells containing the marker might respond differently to a particular stain. It is thus possible, for example, to distinguish between cells that have the marker (these might be female cells) and those that do not have the marker (male cells).

In their initial studies of gynandromorphs, Hotta and Benzer used genetic markers that showed up in the cells that form the wings and the hard outer covering of the insect, the cuticle. They carefully analysed the distribution of male and female cuticle on animals that did or did not perform particular parts of the courtship sequence. From these data they were able to draw inferences about which parts of the nervous system had to be male for particular aspects of courtship to occur. They concluded that the behaviour patterns of following females and wing vibration require that the head, and hence the brain, be made up of male cells; attempted copulation requires both the head (brain) and the thorax (thoracic ganglion) to be male. Close examination of Figure 4.1 reveals that these conclusions are somewhat simplified.

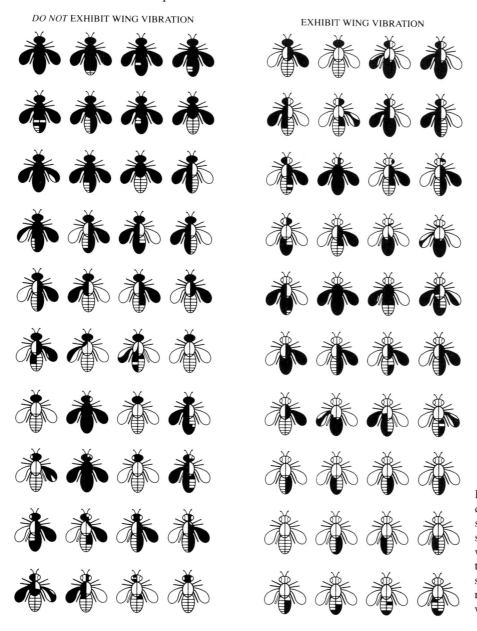

DO NOT EXHIBIT WING VIBRATION

EXHIBIT WING VIBRATION

Figure 4.1   Mosaic fruit flies: female cuticle and wing is indicated by dark shading; white areas are male. The sample of 40 flies shown on the left was chosen at random from the group that did not show wing vibration; the sample on the right was chosen at random from the group that did show wing vibration.

☐ Examine the data on wing vibration presented in Figure 4.1. Do they support or refute Hotta and Benzer's conclusion?

■ By and large they support the conclusion. Most of the flies that did not show wing vibration have predominantly female heads, whereas most of the flies that did show wing vibration have predominantly male heads. However, there are exceptions. There are some wing vibrators that have predominantly female heads, for example. (There were also some non-wing vibrators with male heads, but they did not happen to be selected in the samples and so do not appear in the figure.) It is also far from clear exactly which parts of the head need to be male for wing vibration to occur.

☐ How could such exceptions be accounted for ? (Think about what Hotta and Benzer were observing, and what they were inferring from their observations.)

■ The anomalous results could be accounted for by the fact that Hotta and Benzer could only distinguish between male and female *cuticle*. They inferred that the cell type that made up the cuticle was the same as that which made up the underlying tissue—for example, that the cell type making up the head cuticle was the same as that making up the brain. But it is perfectly possible for a mosaic fly to have male head cuticle with underlying female brain tissue.

Hotta and Benzer's results suggest that this is a rare event and that cuticle type is a fairly good predictor of underlying tissue type. Their results were therefore inconclusive, though they did suggest that the nervous systems of male and female *Drosophila* are different.

A more detailed analysis of the same problem was undertaken by Jeffrey Hall in 1979. He used gynandromorphs and a marker that allowed him to identify each *cell* under the microscope as either male or female. He was able to confirm Hotta and Benzer's conclusion that following behaviour required a male brain, and that attempted copulation required a male thoracic ganglion. However, wing vibration, in contrast to Hotta and Benzer's conclusion, appeared to require only a male thoracic ganglion. Hall's results were extremely detailed, so detailed in fact that he was able to identify a particular small area of the dorsal part of the fly's brain which must be male for following behaviour to occur! The conclusion that there are differences in the nervous systems of male and female *Drosophila* is clear. What those differences actually are awaits further study.

Hall was aware of some subtle problems associated with the use of gynandromorphs. Two of these problems are of general relevance to the study of development. The first is that, where there is a sequence of behaviour, such as in courtship, elements occurring later in the sequence are often dependent on elements present earlier in the sequence being properly performed. Thus, it is only possible to examine the effect of a particular manipulation on later behaviour patterns in a sequence, if that manipulation leaves earlier behaviour patterns in the sequence unaffected. The second problem with gynandromorphs is that tissue with a male genotype has to interact with tissue with a female genotype, and sometimes this is not possible. For instance, it may be that 'male' neurons in one part of the brain form a structure that is absent from the 'female' brain. If the surrounding neuronal tissue has a female genotype, then those 'female' neurons may not interact with the novel male structure in an appropriate way (that is, the target may be inappropriate for the growing axons; see Section 2.4). Thus, the structure could be isolated to a greater or lesser extent from the rest of the brain and nervous system.

A good example of such sexually dimorphic structures in brains is provided by the sphinx moth, *Manduca sexta*.

4.2.2   Sexual dimorphism in the sphinx moth

Many moth species show sexually dimorphic behaviour. In particular, the male will detect female pheromone and fly in a zigzag course towards its source (Book 1, Section 2.7.1), a behaviour pattern not seen in females. The neural basis of this behavioural difference was investigated by Anne Schneiderman and John Hildebrand (1985) at the University of Arizona. They studied the sphinx moth, and found differences between males and females in both the sensory organs on the antennae and in the part of the brain to which the sensory axons from those sensory organs project.

Both female and male antennae have numerous *short sense organs* on them (see Figure 4.2). Each of the sense organs is contacted by sensory neurons, whose axons project via the antennal nerve to a localized region of the brain, the antennal lobe. Within the antennal lobe the sensory axons form synapses with antennal lobe neurons in complex structures called *glomeruli* (singular, glomerulus). (Each glomerulus is essentially a complex arrangement of axons synapsing onto the dendrites of antennal lobe neurons.) Individual sensory axons only project to a few glomeruli, but branch extensively within each. The antennal lobe neurons provide the output from the antennal lobe to the rest of the brain.

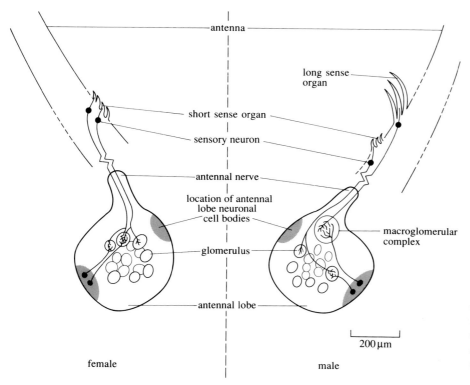

Figure 4.2   Diagrammatic representation of the main features of the neuronal connections between antennal sense organs and the antennal lobe of a sphinx moth. The left side of the diagram shows typical connections in a female; the right side of the diagram shows typical connections in a male. Note the long sense organs and the macroglomerular complex in the male.

81

The male also has *long sense organs* on its antennae in addition to the short sense organs. Furthermore, the axons of the sensory cells that contact the long sensory organs project to an extra-large glomerulus in the antennal lobe, called the *macroglomerular complex*, which is absent in the female. It is the long sense organs that are sensitive to female pheromone.

The important point here is that there are three differences between males and females: a behavioural difference, a difference in their sense organs and a difference in brain structure. Schneiderman and Hildebrand wanted to know how the difference in brain structure came about. They had noticed that, if antennae were absent during development, then the glomeruli were hard to see; they were indiscrete and irregular. Also, if the moths without antennae were male, there was no macroglomerular complex. From this and other observations they put forward the hypothesis that the development of the macroglomerular complex is caused by ingrowing sensory axons from the male antennae. (This suggestion is similar to the experiment of Frank described in the previous chapter (Section 3.3.2), where the peripheral target, muscles, altered the central connections in the spinal cord of the frog (see also Section 3.2.4). Similarly, the development of barrels in the somatosensory cortex of the mouse is dependent on innervation by sensory axons from the whiskers—Section 3.2.4 and Book 2, Section 8.8.2.)

☐   How do you think they could test this hypothesis? (*Note*   The removal of the antennae is an insufficient test.)

■   One way of testing the hypothesis is to have males with female antennal tissue, or females with male antennal tissue—in other words to use gynandromorphs.

☐   If the hypothesis is correct, then would you predict the presence or absence of the macroglomerular complex in:

(a) a male with female antennae;

(b) a female with male antennae?

■   The macroglomerular complex should be absent in (a) and present in (b). Put another way, only where there is a male antenna would you expect to find a macroglomerular complex.

This is precisely what Schneiderman and Hildebrand found. They transplanted antennal tissue at an early stage of development from male into female pupae and vice versa. Development was normal, except that where a male antenna was on an otherwise female moth, a macroglomerular complex was found in the antennal lobe; where a female antenna was on an otherwise male moth, no macroglomerular complex was found in the antennal lobe.

The development of a macroglomerular complex is thus dependent on the innervation of the antennal lobe by the afferent axons of sensory neurons that contact the long sensory organs. Changing the antennae changes the brain, but does the behaviour of the gynandromorphs also change? The behaviour in question has two components: (a) detection of the pheromone, and (b) zigzag flight towards the source of the pheromone (Book 1, Figure 2.15). Using electrodes to record the activity of neurons in the antennal lobe, it is possible to show that, if the antennae are male, these cells change their rate of firing in response to pheromone blown over the antennae. So female moths with male antennae can detect the pheromone.

Zigzag flight towards a pheromone source has been found in some females with male antennae, but in no males with female antennae. The behavioural responses in these gynandromorphs are consistent with the hypothesis that it is the male antennae and their influence on the antennal lobe that determines whether an animal will show zigzag flight.

☐   How would you account for the fact that only some females with male antennae show zigzag flight towards a pheromone source?

■   The most likely explanation is that proposed by Hall to account for the inconsistent behaviour of his *Drosophila* gynandromorphs, namely that the interaction between male and female neuronal tissue is only sometimes successful.

Summary of Section 4.2

Genetic differences between tissues in the same animal do not normally arise. When they do, usually after deliberate experimental manipulation, the resulting genetic mosaic animals reveal something of the influence of the genotype on behaviour. For instance, it has been shown that some aspects of male courtship behaviour require certain parts of the fruit fly to have the male genotype. Different aspects of courtship require different parts of the fly to have the male genotype.

The influence of the genotype on brain structure can be indirect. The formation of the macroglomerular complex in the sphinx moth depends on afferent fibres from long sense organs in the antennae. Long sense organs are normally only found in moths with the male genotype: hence the macroglomerular complex is normally only found in the male. A female moth that develops with a transplanted male antenna not only forms a macroglomerular complex, but also responds to the pheromone released by other females. In this example a particular genotype is necessary for the development of the long sense organ, which in turn is necessary for the development of the macroglomerular complex, which in turn is necessary for a response to the appropriate pheromone.

These examples show that any attempt to explain behaviour in terms of its development must take account of the genotype of the organism. Sometimes, as with the examples presented here, it is possible to identify those structures that must have a particular genotype if one type of behaviour rather than another is to be exhibited by the animal. The crucial point is that environmental factors that influence the development of behaviour do so in conjunction with the genotype.

4.3   External factors and development

The previous section was concerned with the effect of genes on development. This section is concerned with the effect of external factors on development, that is, with the nature of the external stimulation and its effect on the development of behaviour. The principal factors considered here are visual stimuli and sounds, and a group of factors whose mode of action is not entirely clear, called non-specific factors. As many of the effects of these factors are quite subtle, the section begins with an example in which environmental factors cause conspicuous differences between individuals with virtually the same genotype.

4.3.1 Insect castes

In many species of social insects, such as ants, bees and termites, individuals within a colony are specialized to perform particular functions. There are some that lay eggs (queens) and others that fight (soldiers), major workers and minor workers, animals that forage, others that look after the queen and so on. All members that perform a particular set of tasks within the colony constitute a **caste**.

What is striking is that those individuals that constitute a caste often differ from other castes not just in the functions they perform but also in their morphology and physiology (Figure 4.3).

Edward Wilson, who has spent many years studying insect societies, distinguishes between **temporal castes** and **physical castes**. Temporal castes arise where animals perform different tasks at different ages. For example, minor workers of the ant *Pheidole dentata* initially attend the queen and look after the eggs, but as they grow older they look after the larvae and pupae; when older still they perform nest maintenance (for example excavation) and foraging activities (Figure 4.4).

Figure 4.3 Females of the leaf-cutter ant *Atta cephalotes*: a soldier and some smaller workers. Soldiers weigh as much as 90 mg and the smallest workers (not shown) as little as 0.42 mg.

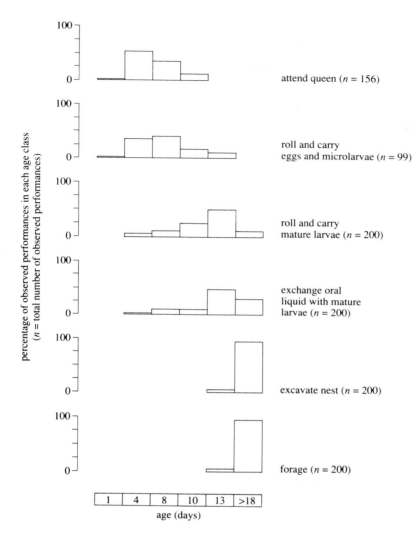

Figure 4.4   Temporal castes in the minor workers of the ant *Pheidole dentata*. The histograms show the percentage of the observed performance of particular behaviour patterns performed by ants of different age groups. The average age of the workers in each group is given at the bottom. (*n* = the total number of observed performances of each task.)

Temporal castes are often associated with physiological differences. For example, the wax glands of the honey bee (*Apis mellifera*) reach their maximum size at about 10 days of age, coincidental with the animal's comb-building activities, and decrease in size thereafter. Their salivary glands, however, are at maximum size at about 20 days of age, at a time when they are generally processing nectar.

Physical castes are like those of the leaf-cutter ant illustrated in Figure 4.3: the external appearance of animals in different castes is different. Sometimes physical castes simply differ in size or proportion, but often there are also morphological (that is, the shape or appendages) differences between the animals in different castes (Figure 4.5, *overleaf*).

Castes are of interest developmentally because they are the result of environmental factors affecting the development of individuals. All the members of a colony are closely related, often being the offspring of a single queen, and often all sisters. The different castes therefore do not have different genotypes. How then could environmental factors account for the different castes?

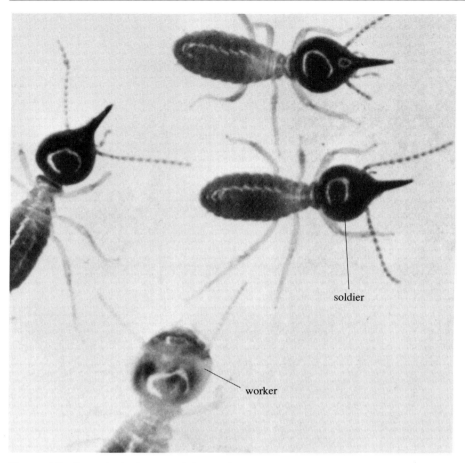

Figure 4.5 Two physical castes of the termite *Nasutitermes exitiosus*. Notice the spout-like nasus (nose) on the head of the soldiers, from which sticky secretions can be sprayed. One worker without the nasus is also shown.

In developmental terms, temporal castes may seem to be easier to explain than physical castes because physiological changes with age are quite common—for example puberty in humans. It could be that the normal processes of development bring about particular physiological states at different ages and that, when an animal finds itself in a particular physiological state, it performs certain behaviour patterns. However, temporal castes are not determined solely by age. For instance, if all the 10-day-old bees are removed from a hive, then the wax glands of some older bees will start to grow again, whereas their salivary glands decrease in size. They are then able to return to a temporal caste they occupied earlier in their lives. Furthermore, the wax glands of some very young bees can enlarge prematurely. Both of these changes can be understood in terms of the functioning of the hive (if there are inadequate numbers of animals building brood combs, then more must be recruited), but not in terms of the mechanisms involved. Quite how the absence of a suitable number of individuals of a particular caste is communicated to other castes and how this information is used to change the other castes is not fully understood.

The development of physical castes presents an altogether different problem: how can such diverse body shapes arise from the same genotype? Wilson identified two major determinants: (a) growth rates and (b) 'decision points'.

(a)   Growth rates

Not all parts of the body grow at the same rate, but the relative growth rates of different parts of the body are very similar among different individuals. Because the relative growth rates are similar, the body forms of most individuals of a species are also similar; for instance, most people's arms are about half as long as their bodies. This is called *isometric growth*. Suppose, though, that the arms of taller people grew proportionately more than those of shorter people. This *non-isometric growth* would result in taller people having arms of, say, three-fifths of their body length and shorter people having arms of, say, two-fifths of their body length. When growth is non-isometric, individuals still have the same structures (arms are still recognizable as arms), but the proportions are altered. Non-isometric growth can account for some castes, such as those of the leaf-cutter ant, where the major differences between castes are the proportions of body parts.

Non-isometric growth is not the whole story though, because some castes have different structures and the size differences still have to be accounted for. Nor does describing the differences between castes reveal the mechanism(s) giving rise to those differences. Factors controlling non-isometric growth are beyond the scope of this course.

(b)   Decision points

Wilson suggested that there are a great number of 'decision points', by which he merely meant external factors that have a specific effect at a specific time. For instance, he identified the size of the egg as a decision point in whether an individual will turn out to be a worker or a queen: the larger the egg the more likely that individual is to become a queen. Only four decision points will be considered here, chosen because they exemplify what other authors have identified as the four main routes by which external factors exert their effects on development.

1   *Larval nutrition*   In all species of social insect the larvae are fed by workers. Which workers feed which larvae and how often they do so affects both the quantity and quality of food received by each larva. A larva that receives a lot of food grows more than one that receives a little. Larval nutrition is an example of a class of developmental determinants called *facilitators*. Facilitators enhance a character or property that is present even in the absence of the facilitator.

2   *Winter chilling*   Although this applies particularly to bees and ants in temperate climes, the principle is a very general one. A part of the nest may become a few degrees colder than another part. Larvae or pupae in that part of the nest may thus get chilled and, as a consequence, be able to respond to a second determinant. For instance, chilled larvae fed a particular nutrient may become queens, whereas unchilled larvae fed the same nutrient may become small workers. Thus, the chilling *predisposes* individuals to respond to a second determinant in a particular way. Predisposition in development is where determinant B only has an effect if event A has happened previously.

3 *Humidity* There may be a slight humidity gradient within the nest. Only larvae developing at a particular humidity can develop into a particular caste. Thus, humidity is an *initiator* of a particular developmental path.

4 *Caste inhibition* Having arrived in a particular caste by dint of nutrition or other determinants, a caste may produce a stimulus (probably chemical) that inhibits further development either in members of its own caste or in members of another caste. A queen may produce a particular pheromone, for instance, that can be detected by other castes and which inhibits development of those other castes into queens. When the pheromone is removed (for example, when the queen dies), so is the inhibition, and other individuals can then develop into queens. In this case the pheromone is acting to *maintain* the developmental stage of the other castes.

The social life of insects may seem somewhat removed from the behaviour of other animals, particularly humans and other mammals. All animals, though, are subject to developmental determinants that facilitate, predispose, initiate and maintain particular developmental paths, and non-isometric growth has been demonstrated in the brain of previously undernourished rats. Furthermore, there are close parallels between the caste system of the social insects and the caste system of a mammal you met earlier in the course. The society of the naked mole rat, *Heterocephalus glaber* (Book 1, Section 10.4.1), consists of a single reproductive female, a number of large soldiers and a lot of smaller workers. If the single reproductive female is removed, one of the soldiers will become fertile and take over the reproductive caste.

4.3.2 Non-specific factors

A number of studies have investigated the effects of non-specific developmental determinants. They are referred to as non-specific because they are ill defined, they are probably multi-modal (that is, involving more than one sensory modality) and it is not clear what their mode of action is. Despite these shortcomings, they remain important because they reveal a number of things about the processes of development and how they should be thought about. So, what are these studies and what do they reveal about the processes of development?

The first group of studies comes under the heading of environmental complexity and is closely associated with the name of Mark Rosenzweig at Berkeley, California. The idea here was that, by raising animals, specifically laboratory rats, in environments that differed in complexity, then the later behaviour of the animals would differ. Young animals were maintained in environments within the laboratory that were defined by the experimenters as complex, or enriched, meaning that they contained other animals of the same age, numerous play objects, running wheels, ladders and alleys. The performance of these animals on later tests of learning and problem-solving was compared with that of animals reared in environments defined by the experimenters as restricted, usually meaning that the animals were kept singly in small uninteresting cages (Book 1, Section 3.6.3).

The behavioural results were by no means consistent, but they did suggest that animals reared in enriched environments were generally better at the learning and problem-solving tasks on which they were tested than those animals reared in restricted environments. Surprisingly consistent results have been reported from studies looking for differences in brain anatomy and biochemistry between animals reared in these two environments. Animals reared in enriched environments have been consistently found to have thicker cerebral cortices, particularly in the occipital area (Book 2, Figure 8.9), and more voluminous dendrites, as measured by length and branching, than those reared in restricted environments. Their brains also have higher levels of acetylcholine. It is not obvious what the relationship is, if any, between the behavioural data and the anatomical and biochemical data.

The second group of studies used an experimental technique called handling. The experimental subject was again the laboratory rat, and the technique involved the experimenter picking up each rat pup in their hand for a few minutes each day. The rats handled each day for the first three or four weeks after birth were compared in adulthood with rats that had not been handled, but which otherwise had had identical treatment.

The behavioural testing of the handled and non-handled animals relied on the **open field**, a large, brightly lit empty arena with no hiding places. There is not much a rat can do in an open field except remain motionless, move about, groom itself, defaecate or urinate. Some of these activities could be easily quantified—length of time spent motionless, amount of movement, number of faecal pellets deposited, etc.—and they have provided the behavioural measures in numerous experiments.

Typical results from such an experiment are presented in Table 4.1.

Table 4.1 Mean activity* of male and female rats briefly handled each day from birth to 25 days of age, or not handled; activity was measured in an open field at 50 days of age.

	Activity	
	Males	Females
non-handled	79	121
handled	228	244

*Activity was measured by the number of 23 × 23 cm squares marked on the floor of the open field that the animal entered during a 3-minute test period.

The handled animals were significantly more active than the non-handled animals. This result has been interpreted to mean that handled animals are less stressed in the open field, and are therefore more likely to move around.

The results of a slightly different experiment carried out by Victor Denenberg and Arthur Whimbey (1963) are shown in Table 4.2. These data show the activity of mature rats whose mothers had been handled or not handled as infants; none of the animals that were tested had received the handling treatment.

Table 4.2 Mean activity* of male and female rats whose mothers had been briefly handled each day from birth to 25 days of age, or not handled. Activity was measured in an open field at 50 days of age. Sixteen animals of each type were tested (that is, a total of 64 animals).

Experience of mothers as infants	Activity of offspring	
	Males	Females
non-handled	76	119
handled	53	59

*Activity score as in Table 4.1

The offspring of handled animals were significantly less active than the offspring of non-handled animals.

☐   What is the biological significance of the data in Table 4.2?

■   That the behaviour of one generation can be influenced by the experience their parents had as infants; the influence is non-genetic, it does not involve the genotype.

This experiment demonstrates *transgenerational* behavioural determinants (Book 1, Section 9.5).

A third group of studies investigated the effects of inadequate nutrition on brain and behaviour in, you've guessed it, the laboratory rat! There have been many investigations of the effects of specific nutritional deficits of, for example, particular minerals, vitamins or amino acids. Such dietary deficits have generally been found to be debilitating and are not considered further. In the studies of interest here, the quantity of food available during early development was reduced while the pups were suckling, but adequate food was provided after weaning. As adults, the brain and behaviour of the previously undernourished animals were compared with control animals that had always been adequately fed.

The previously undernourished animals were always smaller than the controls. Numerous differences in behaviour were found between previously undernourished animals and control animals. The previously undernourished animals were more likely to fight when paired together; they showed greater perseverance of conditioned behaviour under extinction conditions (that is, when reward no longer follows performance of an operant; see Book 1, Section 6.3.3). Perseverance is revealed by the previously undernourished animals continuing to press a lever that used to give a reward long after they should have learnt that pressing the lever no longer produced a reward. They also exhibited exaggerated responses to food and water when deprived (for example by running faster through an alley to a reward). Their brains too were different from those of controls, in that they had a disproportionately small cerebellum and their cortical neurons had fewer dendritic spines.

Although these results may seem unsurprising, and even intuitively reasonable, it is by no means clear that they are all a direct consequence of inadequate nutrition.

☐   In what other, less direct ways, could inadequate nutrition during early life affect behavioural development?

■   Inadequate nutrition could alter the behaviour of mothers looking after well-fed or poorly fed young, with the well-fed young receiving different maternal care than poorly fed young. Also, interactions between siblings could be altered by the nutritional regime. Both of these effects have been found: well-fed young are licked more than poorly fed young, and well-fed young interact with each other more frequently than do poorly fed young.

The final group of studies involved altering the social environment of the developing animal, by removing its peers (animals of the same age) or the mother. Rats, dogs and several primate species have been used in these studies. Some of these studies have already been mentioned in Book 1, Sections 5.3.2 and 9.4. A grossly simplified summary of the main findings from rhesus monkeys (*Macaca mulatta*) is presented in Figure 4.6.

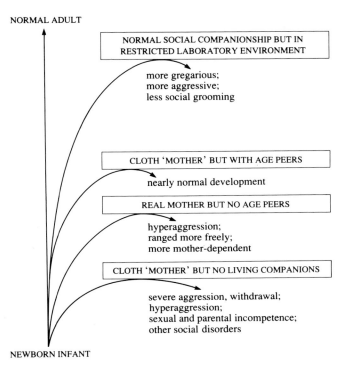

Figure 4.6   Summary of the main findings from studies in which the social environment of the developing rhesus monkey was altered, and the behaviour of the monkey examined at some later time. The effect on the monkey's behaviour compared with a normal adult are indicated.

Essentially the more abnormal the social environment of the developing monkey, the more abnormal its later behaviour, especially its social behaviour.

These disparate groups of studies have revealed a number of important things about developmental processes. The three most important of these are:

1   that the determinants are interrelated; that is, altering one determinant probably alters another;

2   that it is difficult to distinguish between developmental determinants;

3   that changes in the environment of the developing organism can produce detectable changes in brain anatomy and biochemistry.

Each of the manipulations discussed above is, from the experimenter's perspective, just one manipulation, be it altering the complexity of the environment, or altering the amount of food available. From the animals' perspective, each manipulation introduces a number of changes; a complex environment means the animals get more exercise, can escape from social companions, can move in three dimensions and so on. Any or all of these changes can be thought of as a developmental determinant and they are all interrelated; alter one factor and something else also changes. Given this interrelatedness, it becomes very difficult to establish that the developmental determinant that the experimenter has manipulated is the direct and specific cause of some behavioural difference, rather than causing a change in some second or third determinant which then causes the behavioural difference.

The significance of the differences in brain anatomy has yet to be established; they are present, but it is not clear how they relate to the environmental manipulation or to any behavioural differences. However, it is clear that the changes are long-lasting and may alter the animals' behavioural potential in appropriate circumstances.

These studies throw up two other things that are worthy of note. Firstly, all of the manipulations discussed generate variation. For an ethologist, this is all-important because any comparable changes in a natural context would lead to variation on which natural selection can act. A psychologist, though, has an additional interest, for it is also important to know in which direction the determinant alters behaviour. If a complex environment early in life improves the later problem-solving ability of rats, might it also improve the potential of human infants? If early handling reduces the potential for stress in later life, might this be applied to human development? Answers to such intriguing questions must await further study. Secondly, it is sometimes difficult to establish whether a developmental determinant has any effect on behaviour. Much as the problems of detecting what an animal has learnt rest on procedures for testing the animals, so the effect of a developmental determinant on a particular behaviour pattern may be missed if an appropriate test is not used or if it is assumed that only a small range of behaviour patterns will be affected.

4.3.3 Modality-specific factors

The role of two modality-specific factors—visual stimuli and sounds—in shaping the nervous system and behaviour is considered in this section.

Visual stimuli

The final determination of neuronal connections within the mammalian visual cortex occurs to a considerable extent post-natally and, hence, at a time when the nervous system is under the influence of visual experience. In view of the fact that neuronal activity plays a role in developmental processes (Section 3.2.3), it follows that the kind of visual experience an animal has in early life might affect those neuronal connections and hence its visual abilities.

Two sets of experiments will be described: competition for ocular dominance and reverse-occlusion.

Many cells in the visual cortex respond better to stimulation of one eye than the other. This is the phenomenon of *ocular dominance*. In studies of competition for ocular dominance, changes are found in the fine structure of the cortex as a result of restricting the vision of one eye by closing it (occlusion) while allowing normal vision in the other eye. The experiments have used a number of different techniques, including electrophysiological recording of single neurons in the cortex (Book 2, Box 8.3) and retrograde labelling (Box 4.1).

Box 4.1   Retrograde labelling of the visual cortex

Some substances are taken up by presynaptic terminals and distributed throughout the neuron. They may subsequently be released into synaptic clefts and taken up by other presynaptic terminals. They are therefore transported along axonal pathways. One such substance is the amino acid proline. If proline is labelled with radioactive hydrogen and injected into one eye, it will eventually be found in the visual cortex. The exact location of the proline in the visual cortex can be determined using autoradiography. Any part of the cortex containing the proline must have a connection with the eye originally injected. Studies using proline have revealed extensive banding in the visual cortex, representing the ocular dominance bands.

The various techniques have yielded similar results, so the principal discoveries from these studies are discussed in terms of the technique introduced in Book 3, Section 4.3.4, of injecting the sugar-like substance, 2-deoxyglucose (2-DG). Briefly, 2-DG is injected into the bloodstream and is taken up by neurons as if it were glucose. However, unlike glucose, 2-DG cannot be utilized as an energy source, and its breakdown products accumulate in the neuron. The most active neurons require the greatest energy supply and thus accumulate most 2-DG. If 2-DG is labelled with a radioactive marker, then, when the brain is cut into thin sections and each section is placed on a photographic emulsion, radiation from the labelled 2-DG exposes the emulsion. (This is the technique of autoradiography— Book 3, Box 4.1.) When the 'photograph' is developed, a picture of the pattern of radiolabel results, and reveals areas of the tissue that were utilizing glucose, and hence were active.

If you were to use this technique to measure activity in cortical neurons in normal animals responding to stimulation of one eye with light, you would see bands; areas of high 2-DG accumulation would alternate with areas of no 2-DG accumulation, and in between would be areas of intermediate 2-DG accumulation (Figure 4.7; Book 3, Section 4.6.1, Figures 3.48 and 3.49).

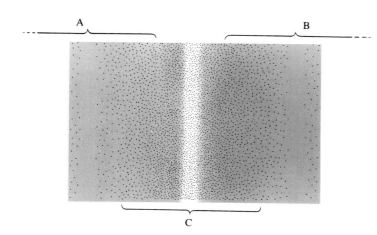

Figure 4.7   Schematic section through a small part of the visual cortex showing bands of 2-DG accumulation. Bands A and B show areas where inputs from one eye terminate. Band C shows an area where inputs from the other eye terminate. Stimulation of one eye would reveal activity in bands A and B; stimulation of the other eye would reveal activity in band C. Note that band C overlaps with bands A and B; that is, some cells receive inputs from both eyes.

☐ In terms of inputs from the stimulated eye to the cortex, what do these bands represent?

■ The areas of high 2-DG accumulation represent areas of high neural activity. The high activity is in response to the visual stimulation. Therefore these areas correspond to those cells that are 'driven' predominantly by (that is, receive inputs predominantly from) the stimulated eye. The areas of no 2-DG accumulation correspond to those cells driven predominantly by the non-stimulated eye. The areas of intermediate accumulation correspond to those cells normally driven by both eyes (binocular cells). To show high levels of 2-DG accumulation in binocular cells, they need to be driven simultaneously by both eyes (Figure 4.7).

The extent to which a cortical cell is driven by a particular eye is its *ocular dominance*. It is measured on a scale of 1 to 7; binocular cells are classes 3, 4 and 5; cells driven predominantly by one eye are in classes 1 and 2; cells driven predominantly by the other eye are in classes 6 and 7 (Figure 4.8).

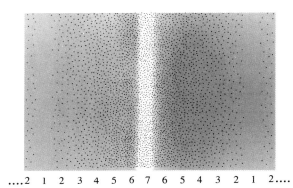

....2 1 2 3 4 5 6 7 6 5 4 3 2 1 2....

Figure 4.8 The same small section of visual cortex as shown in Figure 4.7, but with regions numbered on the ocular dominance scale according to whether cells in those regions are predominantly driven by one eye, the other eye or both eyes.

The 2-DG procedure can be used to examine the visual cortices of rhesus monkeys in which one eye had been occluded, and therefore deprived of normal visual experience, for the first three months after birth (Figure 4.9a). After this time the deprived eye of the animal was stimulated with light while 2-DG was administered via the bloodstream. Distribution of 2-DG in the cortex revealed that there was very little incorporation of 2-DG resulting from stimulation of the deprived eye (Figure 4.9b).

☐ What does this tell you about the effect of the visual deprivation?

■ Visual deprivation must have somehow reduced the effectiveness of inputs from the deprived eye to the cells within the visual cortex. The reduced effectiveness meant that subsequent stimulation of the deprived eye failed to elicit activity in the visual cortex. As there was no activity, no 2-DG was taken up into cortical cells.

☐ What would you predict about the pattern of 2-DG accumulation in the cortex if the experiment were repeated, but with the non-deprived eye stimulated?

■ You might predict that there should be accumulation of 2-DG in cells lying within bands in the cortex where inputs from the non-deprived eye project.

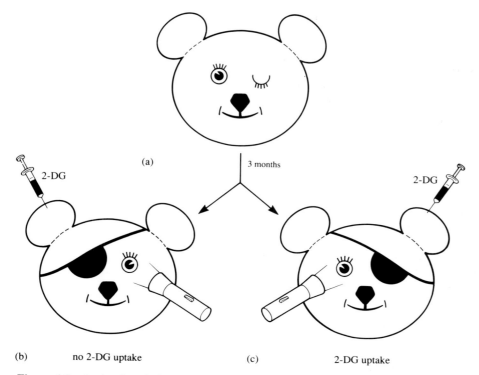

(a)

3 months

2-DG

2-DG

(b)   no 2-DG uptake   (c)   2-DG uptake

Figure 4.9 A visual occlusion experiment on rhesus monkeys: (a) closure of the left eye from birth until three months of age; (b) stimulation of the previously deprived eye with light does not result in the uptake of 2-DG in the visual cortex; (c) stimulation of the non-deprived eye results in the uptake of 2-DG in the visual cortex. (b) and (c) demonstrate that cortical cells are in ocular dominance classes 1 and 2 (or 6 and 7) only.

There was uptake of 2-DG (Figure 4.9c), but there was no evidence for any banding. The visual cortex had a uniformly dense pattern of 2-DG accumulation. The lack of banding is surprising and suggests that axon terminations that carry the input from the non-deprived eye now contact the cells that would normally receive input from the deprived eye. (An alternative explanation in terms of the selective death of cells that normally receive input from the deprived eye is incorrect because eye closure does not lead to a loss of cells in the visual cortex.) The change in projection occurs possibly by an enlargement of the area in which synaptic terminations from the non-deprived eye are made. In turn, this suggests that collateral sprouting of axons (Section 3.6.1) from the non-deprived eye occurs, leading to the formation of new synapses on neurons that, at birth, received terminations from the deprived eye. The new synapses take over and occupy space previously used by synapses from the deprived eye. These latter synapses are presumably weakened because during the three-months period of occlusion they received little or no stimulation.

Similar results to those obtained in rhesus monkeys have also been obtained in domestic cats, *Felis catus*.

Monocular occlusion, then, results in the absence of cortical cells driven by the deprived eye and the absence of binocularly driven cortical cells in the visual cortex; all cells are driven exclusively by the non-deprived eye. A very similar effect of visual stimulation on function is seen when stimulus orientation is manipulated, as now described.

Kittens raised for the first three months of life in a visual environment consisting solely of vertical lines are subsequently unable to see horizontal lines, and they bump into horizontal bars. Similarly, kittens raised for the first three months of life in a visual environment consisting solely of horizontal lines are subsequently unable to see vertical lines, and bump into vertical bars. Neurophysiological examination of the cortices of cats reared under such strange visual conditions reveals a mechanism similar to that described above for ocular dominance shifts. Just as cells in the visual cortex show differential responses to stimulation from each eye (that is, ocular dominance), so they show differential responses to the *orientation* of the stimulus (Book 3, Section 4.3.4 and Figure 4.28). In other words, cells in the visual cortex respond to some stimulus orientations and not others, whereas other cells respond to different stimulus orientations. When kittens are raised under conditions of a single stimulus orientation, axon terminations from cells that carry information about that stimulus orientation make contact with cells that would normally receive input from cells carrying information about different stimulus orientations. Thus, cells in the visual cortex come to be responsive only to one stimulus orientation, in much the same way as cells come to be responsive only to the non-occluded eye.

Does it matter *when* the visual input is altered for these changes in the visual cortex to occur? In other words, is there a sensitive period in the development of the visual system? This was investigated by so-called 'reverse-occlusion' experiments. In such experiments, the time-course of the visual system's susceptibility to monocular occlusion was studied by first closing one eye shortly after birth, then, after a certain period of time, opening it and closing the other eye.

The effect of the first occlusion is to shift the ocular dominance of cortical cells in favour of the open eye, so that, for instance, classes 1 and 2 dominate (Figure 4.8). There are few cells in classes 3, 4, 5, 6 and 7. Following reverse-occlusion the ocular dominance distribution is shifted back in favour of the eye that was initially closed but then opened. Thus, there is a decrease in the number of cells in classes 1 and 2, and an increase in cells in classes 6 and 7 (or vice versa). The bands of termination corresponding to the newly opened eye will be expected to enlarge to take over silent cells previously driven by the eye that was closed at reverse-occlusion. Reverse-occlusion done within one and a half months restores the normal balance of cells in classes 1–7. A summary of these events for the cat is presented in Figure 4.10.

If synaptic connections in the cortex are plastic (that is, they can be altered) only for a limited period following birth, then an increase in the length of time before reverse-occlusion takes place should result in a reduced effect; that is, the longer an eye is closed before reverse-occlusion, the less the recovery of the banding from that eye should be. Behavioural, electrophysiological and anatomical studies of reverse-occlusion experiments have all tended to produce similar results. Cats that have undergone reverse-occlusion *early* in life recover vision in the initially closed eye; the distribution of ocular dominance shifts back in favour of the initially closed eye. Reverse-occlusion produces an effect on the functional state of

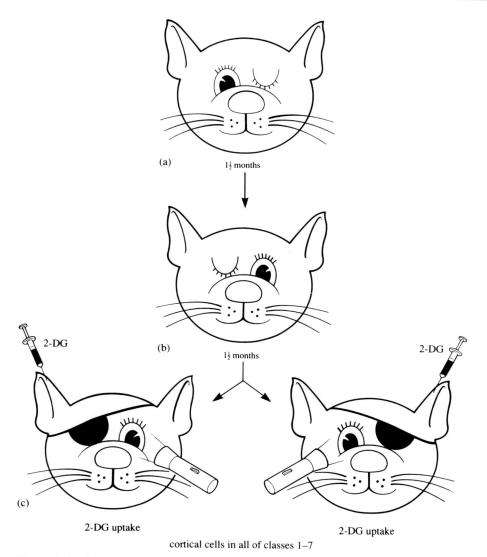

Figure 4.10   Summary of the effects of one *early* reverse-occlusion experiment in the cat: (a) monocular occlusion of the left eye; (b) reverse occlusion; (c) testing for ocular dominance.

the visual cortex if it occurs within the first two or three months after birth. The later that reverse-occlusion occurs within this period, the smaller are the changes. If reverse-occlusion occurs after three months, ocular dominance does not shift back in favour of the initially closed eye (Figure 4.11, *overleaf*).

Monkeys also show a sensitive period to reverse-occlusion, but it is longer than in the cat (about a year). Likewise, in human infants, unless a squint is corrected early on, later visual acuity in the squinting eye is impaired, and the children do not have binocular vision.

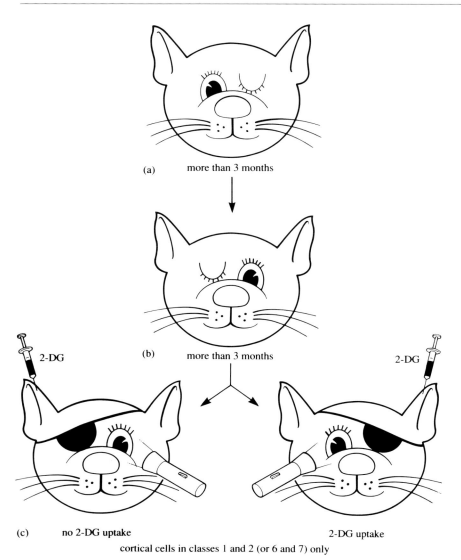

(a) more than 3 months

(b) more than 3 months

2-DG

2-DG

(c) no 2-DG uptake   2-DG uptake

cortical cells in classes 1 and 2 (or 6 and 7) only

Figure 4.11 Summary of the effects of a *late* reverse-occlusion experiment in the cat: (a) monocular occlusion of the left eye; (b) reverse occlusion; (c) testing for ocular dominance.

Thus, a sensitive period in visual cortex development seems to be common in mammals. Reverse-occlusion experiments indicate a period in visual cortex development (as described above for the kitten) when its fine synaptic structure is plastic. At this stage, the visual system is able to fine-tune to the actual positions of the eyes, or indeed to compensate for changes in input from either eye (Figure 4.10). For example, if one eye were damaged during this time, then the other would be able to strengthen the synaptic effectiveness of its inputs at the expense of the weakened inputs from the damaged eye. Changes in inputs after the sensitive period produce no compensation (Figure 4.11).

These studies suggest that visual experience plays a crucial role in shaping the fine structure of the visual system, a conclusion similar to that of Shatz and Stryker (Section 3.4.2). You may recall that they concluded that 'spontaneous action potential conduction was important in shaping the segregation of the retinal afferents'.

Sound localization

Barn owls (*Tyto alba*) use hearing as well as vision to detect and localize prey.

☐ Which part of the brain is primarily involved in the analysis of visual and auditory information in birds, and how is it organized?

■ In birds, the optic tectum plays an important role in the analysis of both auditory and visual information. The projections of neurons from these two modalities are arranged so that a single point in space is represented at a single point in the tectum for both modalities.

Eric Knudsen in California set out to discover how this convergence of input developed. Barn owls are experts at sound localization, and are capable of locating sounds to within one or two degrees in both the horizontal and vertical dimensions. They are altricial (Book 1, Section 5.1) and grow very quickly, the skull doubling in size in the first seven weeks of life. They hear sounds much as humans do, being sensitive to the same range of audible frequencies, and they do not locate sounds by moving their ears. An unusual feature of the barn owl's ears is that they are asymmetrical: they are not in exactly the same position on opposite sides of the head. The result is that the left ear is more sensitive to sounds from the left and below, whereas the right ear is more sensitive to sounds from the right and above.

The ability of barn owls to pinpoint sound sources indicates that these animals associate a particular balance of acoustic cues with a particular location in space. This ability comes about in much the same general way that visual abilities come about. Most of the necessary neuronal connections are made before hatching, and these connections are then modified by early sensory, in this case auditory, experience.

The influence of experience on sound localization was first demonstrated by placing an ear-plug in one ear—monaural occlusion (Knudsen, 1988). An owl that is monaurally occluded mislocalizes sound sources in a predictable manner: to the left and down from the actual location when the right ear is occluded, and to the right and up when the left ear is occluded. If an owl is less than about 60 days old when one ear is occluded, it first of all makes mistakes, then learns to interpret correctly the abnormal cues induced by the ear-plug; its accuracy of sound localization gradually recovers over a period of weeks (Figure 4.12, *overleaf*). If the owl is older than 60 days when the ear is first occluded, no adjustment of sound localization occurs; the animal maintains a sound localization error for months but, when the ear-plug is removed, it locates sounds with normal or near-normal accuracy immediately. Thus, the auditory system of the young owl is similar to the visual system of the kitten, in that both can adjust to the effects of occlusion up to a certain age, but adjustment is not possible after that age. There would appear to be a sensitive period in this system, just as there is in the visual system.

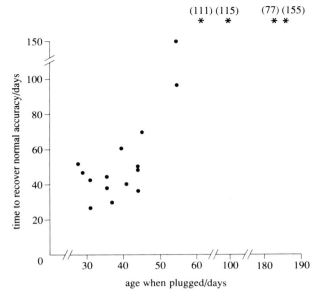

Figure 4.12 The sensitive period for adjusting sound localization accuracy in barn owls, as revealed by monaural occlusion. The number of days required after the owl's ear has been plugged for it to recover normal localization accuracy is plotted against the age of the owl at the time the ear was plugged. A star indicates that no recovery occurred, even though testing continued for a considerable period of time; the number of days for which testing was continued after the ear-plug had been removed is indicated in parentheses.

What happens when the ear-plugs of those owls that did adjust their sound localization were removed? The answer depends on when the ear-plugs are removed (Figure 4.13). Owls younger than 150 days old when the ear-plug was removed adjusted within a few weeks (owls 1 and 2 in Figure 4.13). Owls older than this were slower to recover, and those over 200 days old when the ear-plugs were removed never recovered accurate localization (owls 4 and 5 in Figure 4.13).

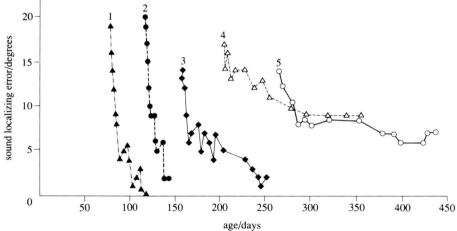

Figure 4.13 Recovery of sound localization accuracy in barn owls after removal of monaural ear-plugs. Data are shown for five representative owls (1–5). All animals had one ear occluded before 40 days of age. The rate of recovery following ear-plug removal slows with age, and adjustment virtually ceases beyond about 290 days of age.

In the initial occlusion experiment, the owls older than 60 days could not adjust to the ear-plug. However, when the ear-plug was removed from these owls, adjustment was possible up to about 200 days.

☐ What do the differences in the owls' abilities to adjust to the initial occlusion and the subsequent removal of the occlusion suggest?

■ It suggests that adjustment is more likely (easier) when both ears are stimulated (following subsequent removal of the occlusion) than when only one ear is stimulated (during the initial occlusion). In other words, the plasticity of the localization mechanism favours associations based on normal auditory stimulation.

One further piece of evidence provided by Knudsen is that sound localization is dependent on normal vision. An owl that grows up seeing the world through prisms that shift its visual world to the left or right adjusts its sound localization to match. For example, an animal raised with prisms that shift vision by 10° to the right responds to sounds by facing 10° to the right of the source.

☐   What does this tell you about the relationship between sound localization and visual localization of the same stimulus?

■   Sound localization in the owl is not an absolute measure, but adjusts to fit in with visual space. In other words, visual space organizes auditory space.

Despite the title of this section, it should be clear that to identify a factor in one modality and assume that it has effects on its own and in isolation from factors in other modalities is too simplistic. However, the effect is perhaps easier to understand when only one modality is specifically altered.

Summary of Section 4.3

The unifying theme of this section has been the effect of external factors on the development of behaviour. It began by considering how castes could arise within a genetically homogeneous colony of ants. The ideas of non-isometric growth and decision points were introduced. Various non-specific factors, including handling, environmental complexity and social environment, influence development in subtle ways, the effects sometimes being transgenerational. Finally, this section looked at the modality-specific factors of visual stimuli and sound. In cats and monkeys, the visual cortex exhibits considerable ability to adjust to monocular instead of binocular input. This ability is only retained for the first few months of life and for a shorter period in cats than in monkeys. Similarly, the auditory system of owls is only flexible early in life.

4.4   Hormones and development

The study of the effect of hormones (Book 1, Section 2.9) on development may at first seem to be relatively straightforward. The source of the hormone can be removed by removing the appropriate endocrine gland, and the hormone can be replaced by injection: these are the standard techniques of endocrinology. There are, though, two particular problems with this approach of which you need to be aware: one has been known for a long time and one has emerged only relatively recently.

The first problem has to do with homeostasis, a concept you met in Book 1, Section 7.2.2. The overall hormonal balance of an organism is the result of numerous feedback loops between hormones, with many hormones affecting the production of both their own and of other hormones. Thus, removal of one hormone can alter the overall balance of hormones, just as removing one member of a hockey team alters the balance of the team. The other hormones—the other members of the team—adjust to the new situation and a new 'balance' is established. Thus, although it is easy to ascribe any differences between experimental and control animals to the missing or injected hormone, such differences may be effected by the altered overall hormonal balance.

The second problem is that many hormones appear to be secreted in pulses, rather than as a continuous flow: they are secreted more at certain times of day than at others. This makes monitoring the level of a hormone in the blood difficult, because the level fluctuates with the timing of the pulses. An injection is a bit like a pulse, but injecting the hormone in a way that mimics normal secretion—that is, in a truly physiological way—is not straightforward. A large dose of hormone given at one point in time, the usual technique, is not the same as several smaller doses given over a longer period of time, even though the total amount of hormone given may be the same.

One final point: species can differ in the hormones they produce, in the precise structure of particular hormones, in when they produce them, and in the effect of those hormones on behaviour and development.

4.4.1 Metamorphosis

For most species, development is a process of continuous and gradual change. Some species punctuate their development with one or two dramatic changes in structure (for example, when a tadpole changes into a frog), the process known as metamorphosis. Metamorphosis allows animals to exist in more than one specialized form and thereby perform activities required for survival and reproduction in different environments, utilizing different resources in those environments. The animal undergoes substantial reorganization during metamorphosis; in the case of the tadpole it loses gills and grows lungs, changes its digestive system and its method of feeding, loses its tail, grows legs and changes its method of locomotion. There are also considerable changes in the nervous system. For instance, the tadpole has particular sense organs that form the lateral line along each side of its body, whereas the adult frog does not have these sense organs; also the control of swimming by the nervous system of the tadpole is different from the control of jumping in the frog. All these changes are under the control of two hormones; one, thyroxine, promotes metamorphosis, and the other, prolactin, inhibits metamorphosis. A bullfrog tadpole (*Rana catesbeiana*) deprived of thyroxine remains a tadpole and continues to grow; if the thyroxine is replaced, it metamorphoses into a bullfrog. Conversely, injections of prolactin during metamorphosis arrest metamorphosis.

The reorganization of the nervous system raises a number of questions about how it is achieved. Do neurons die to be replaced by new neurons? Do the same neurons simply reorganize their connections, or do new neurons arise to perform the new functions, leaving old neurons redundant? Each of these processes occurs to some extent, but this section focuses on one particular set of sensory neurons in the sphinx moth and the hormonal control of their reorganization during metamorphosis.

Metamorphosis in insects, like that of the amphibians discussed above, is controlled by two key hormones. In the case of insects, these hormones are ecdysteroid and juvenile hormone (these hormones affect metamorphosis in the same way as thyroxine and prolactin, respectively, in the tadpole). Richard Levine in Cambridge has studied the effect of juvenile hormone on a small group of sensory cells called the *gin-trap*.

The sphinx moth, like other moths and butterflies, has a larval stage and a pupal stage before adulthood (Figure 4.14).

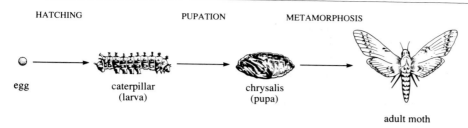

HATCHING   PUPATION   METAMORPHOSIS

egg   caterpillar   chrysalis   adult moth
             (larva)   (pupa)

Figure 4.14   Stages in the life of a moth.

Near the end of larval life, most of the sensory hairs that cover the body of the caterpillar disappear, and their associated sensory neurons die. However, small groups of these sensory hairs, in characteristic locations, remain to form specialized pupal structures—the gin-traps. The sensory neurons associated with these remaining sensory hairs also remain (Figure 4.15). Gin-traps are found paired on opposite sides of the pupa at the rostral end of segments 5, 6 and 7. (There are usually about 13 segments altogether.) Once the gin-traps are formed, movement of their sensory hairs evokes a stereotyped reflex response. This pupal reflex consists of a rapid contraction of the body-wall muscles in an adjacent segment on the side that was stimulated. The important point to remember is that the sensory neurons involved in this response are present in the larva, but the larva does not have the reflex.

What Levine and his co-workers discovered is that, when the moth changes from larva to pupa, the axons of the surviving sensory neurons grow by increasing their length and by producing more branches, and that this growth is necessary if the reflex is to be shown (Levine, 1986). The growth occurs during the final four

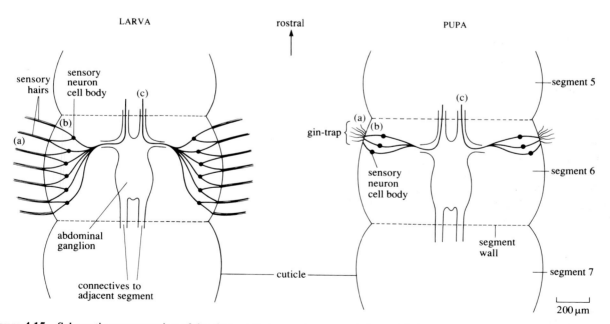

Figure 4.15   Schematic representation of the changes that occur to sensory hairs and their sensory neurons in the sphinx moth when a larva (left) changes into a pupa (right). *Note*   (a) differences in the distribution of sensory hairs; (b) differences in the number of sensory neurons; (c) the projection of the sensory neurons through the abdominal ganglion to an adjacent segment. Abdominal ganglia, sensory neurons and sensory hairs are shown for segment 6 only.

days of larval life, and is controlled by the hormonal environment. The effect on the sensory neurons of raising the level of juvenile hormone at this time is shown in Figure 4.16.

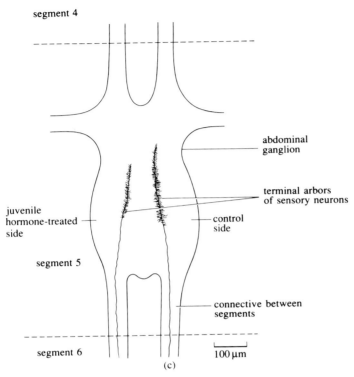

Figure 4.16 Diagram showing the outline of the abdominal ganglion in segment 5 of a sphinx moth pupa. The arbours of two sensory neurons that innervate gin-traps are shown. The cell body of the neuron on the left was treated with juvenile hormone 4–5 days before pupation; the one on the right was untreated. The cell bodies of the sensory neurons are in segment 6, close to the gin-traps. Sensory neuron (c) projects to the gin-trap in segment 6 (see Figure 4.15).

The treated sensory neurons retained their small larval branching pattern and although they were responsive to touch, they failed to evoke the pupal gin-trap reflex response. The untreated sensory neuron in the same segment of the same pupa gave rise to normal pupal arbours and evoked the gin-trap response when touched.

☐ What would you predict about the level of juvenile hormone in the larva prior to pupation stage?

■ The level of juvenile hormone should be low. The experimental results depicted in Figure 4.16 show that, if the level of juvenile hormone were to remain high, then the arbours would remain larval.

The level of juvenile hormone is indeed low in the last few days before pupation. This glimpse at metamorphosis reveals that the hormonal environment is crucial to the reorganization of neurons, to changes in behaviour and to changes in body form that occur at this time.

4.4.2 Sex differences

In many species there are differences in behaviour between the sexes. Sometimes the distinction is absolute, as in the case of bird song, where frequently it is only the male that sings (see Section 4.4.3). More usually the differences between the sexes are of degree rather than kind. For instance, in sexual behaviour, female rats do not usually mount male rats, but sometimes they do; equally, males do not usually adopt the lordosis posture (Book 1, Figure 7.3) but sometimes they do. Other aspects of differences in sexual behaviour could be dependent on the presence or absence of genital structures. Thus, a female may be unable to penetrate because of the absence of a penis rather than because of the inability to make appropriate pelvic movements or the absence of appropriate motivation.

Unlike sexual behaviour and song, social play is present very early in life, yet, here too there are differences between the sexes. In the Norway rat (*Rattus norvegicus*), play-fighting is a behavioural sequence, which begins when one animal pounces on another. The pounce is followed by wrestling and/or boxing and the play-fight usually finishes with one animal on top of the other. A similar sequence of play-fighting is seen in rhesus monkeys. Throughout adolescence the males of both species initiate and become involved in play-fights more frequently than females. The females do play-fight, and the behavioural components of their play-fighting are the same as those used by males, but the female adolescents play-fight less frequently.

So how do these differences in play-fighting and sexual behaviour arise? Part of the answer appears to be due to testosterone. If this hormone is administered to female rats at about the time of birth (neonatal rats), they engage in much more play-fighting as adolescents (between P26 and P40) than normal females. Furthermore, if male rats are castrated just after birth, they engage in play-fighting at a typical female rate. Testosterone injections into juvenile females, or castration of males later than about the sixth day after birth, do not affect the frequency of play-fighting. It would seem therefore that there is a sensitive period from birth to P6 days for responsiveness to testosterone in the rat. These results for the rat are very similar to those found for the rhesus monkey, except that the procedures need to be carried out pre-natally (that is, any time within a month before birth) in order to alter the frequency of social play in the adolescent monkey. Post-natal treatment (that is, within a month after birth) is without any effect. These results are summarized in Table 4.3 .

Table 4.3   A summary of the results of investigations into the hormonal basis of sex differences in play-fighting. 'Male-like' play-fighting refers to high levels of play-fighting; 'female-like' refers to low levels of play-fighting.

Norway rat

Genetic sex	Neonatal treatment (P0–P6 days)	Post-natal treatment (P10–P20 days)	Social play
female	testosterone injection		male-like
male	castration		female-like
male		castration	male-like

Rhesus monkey

Genetic sex	Pre-natal treatment	Post-natal treatment (P0–P1 month)	Social play
female	testosterone injection		male-like
female		testosterone injection	female-like
male		castration	male-like

These results provide evidence that testosterone has an *organizing* effect (Book 1, Section 5.3.1); that is, neural structures that underlie play motivation are organized under the influence of testosterone. You should recall from Book 1 that hormones also have *activational* effects.

☐ What evidence do the results in Table 4.3 present about an activational effect of testosterone on play-fighting?

■ Post-natal castration, and thus the absence of testosterone later in life, does not affect the frequency of play-fighting shown by males. Thus, testosterone does not need to be present for the male-like frequency of play-fighting to be shown. Therefore testosterone does not activate play-fighting.

The pattern for sexual behaviour is similar to that for play-fighting, except that in this case male-like behaviour requires the activating effect of testosterone as well as its organizing effect.

As a final example of the effect that hormones during development have on the differences in behaviour between males and females, consider the Japanese quail (*Coturnix japonica*). Adult male and adult female Japanese quail differ in their tendency to perform such behaviour patterns as copulatory mounting, strutting and crowing. Males are more likely to perform these behaviour patterns in response to appropriate stimulation than are females (Figure 4.17, male and female controls). The effect on these three behaviour patterns of treating males and females with testosterone before hatching (that is, while still in the egg) is shown in Figure 4.17.

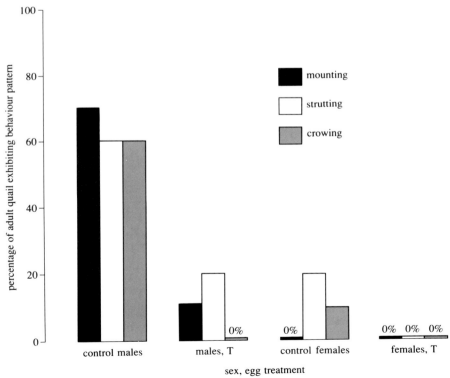

Figure 4.17 The effect of treating quail eggs with testosterone on adult reproductive behaviour; T = testosterone injection.

☐   What effect did testosterone have on the behaviour of the males?

■   It prevented crowing, and greatly reduced the incidence of strutting and copu-latory mounting.

☐   What effect did testosterone have on the behaviour of the females?

■   It prevented strutting and crowing. However, because few females normally exhibit those behaviour patterns, the effect of the testosterone treatment on the females was less dramatic than the effect on the males.

☐   Expressed in terms of changes in behaviour towards male-like or female-like behaviour, did neonatal testosterone administration have the same effect in quail as in the rat and monkey?

■   No. Injection of testosterone into male quail eggs makes the adults more female-like, whereas an injection of testosterone into female rats and monkeys makes them more male-like. Put another way, the quail embryo develops into a male unless testosterone is present, and the monkey and rat embryos/neonates develop into females unless testosterone is present.

During development the hormone testosterone has a powerful organizing effect on subsequent behaviour. Furthermore, the effect can be seen in both male and female neonates; in other words both the male and female are sensitive to testosterone, and respond to it in the same way. You will not be surprised to learn that both sexes have testosterone receptors.

☐   Propose a simple hypothesis which could account for the differences in play-fighting between normal male and female rats and monkeys (summarized in Table 4.3).

■   One hypothesis might be that male neonates have a source of testosterone which exerts its organizing effect. Females do not have a source of testos-terone.

This simple hypothesis is more or less correct. Testosterone is secreted by the male testis from a very early age. Female neonates do not have testes and produce negligible amounts of testosterone.

☐   Does this interpretation fit the quail results shown in Figure 4.17?

■   No. It is exactly the opposite. If female quail do not produce testosterone, then their behaviour should be like that of the control males in Figure 4.17; yet it is not. Similarly, if male quail have a source of testosterone, then their behaviour would be like that of the female quail that had been treated with testosterone; yet it is not.

It turns out that the results presented in Figure 4.17 can equally well be achieved by injecting oestrogen instead of testosterone. Oestrogen is secreted by the ovaries of neonatal female quails. Thus, in the quail, the female produces a hormone that renders its behaviour female-like; the male produces very little of that hormone and its behaviour is male-like.

This leaves two questions unanswered: (a) do pre-natal male quail produce testosterone, and (b) how can two different hormones, oestrogen and testosterone, produce the same effects? The answer to the first question appears to be that prenatal male quail *do* produce testosterone, but it is broken down very rapidly before it can exert its effect. The answer to the second question is a bit more complicated.

Oestrogen and testosterone are both steroid hormones, though for the present discussion their chemical nature is not important. Neither hormone is stored, but they are used as they are made. Both hormones are initially secreted into the bloodstream, from which they can easily pass into cells; unlike neurotransmitters and peptides, no special receptors are needed on the cell membrane for steroid hormones. There are, however, special receptors for oestrogen and testosterone *inside* cells, and the combined receptor/hormone acts directly on the cell's DNA, which then produces particular proteins. (The testosterone/receptor complex produces different proteins from the oestrogen/receptor complex.) The important point is that once inside a cell, testosterone can be readily converted into other hormones, including oestrogen and dihydrotestosterone, provided only that the cell contains the appropriate enzymes. Note that dihydrotestosterone cannot be converted into oestrogen. The actions and conversions of testosterone are summarized in Figure 4.18.

☐   Use the information presented in Figure 4.18 to answer question (b) above about how neonatal testosterone and neonatal oestrogen can have the same effect on male quail behaviour.

■   The active hormone within the cell must be oestrogen. If testosterone is administered, it can be converted in the cell to oestrogen, and the oestrogen/receptor complex will act on the cell's DNA to produce protein 3.

The frequency of play-fighting in female rats and rhesus monkeys treated with testosterone or dihydrotestosterone, but not oestrogen, in early life is male-like. The frequency and appearance of mounting (a component of sexual behaviour) is male-like in female rats treated with testosterone or oestrogen early in life, but dihydrotestosterone has no effect. The observations are summarized in Table 4.4. The inference drawn from these experiments was that some organizing effects of testosterone are the result of testosterone being converted to oestrogen, whereas others require testosterone itself or dihydrotestosterone.

Table 4.4   The effect on the relative frequency of adult play-fighting of treating female rats and rhesus monkeys with different hormones; also shown is the relative frequency of mounting in female rats.

Neonatal hormone treatment	Frequency of play-fighting (rats and rhesus monkeys)	Frequency of mounting (female rats)
testosterone	male-like	male-like
dihydrotestosterone	male-like	female-like
oestrogen	female-like	male-like

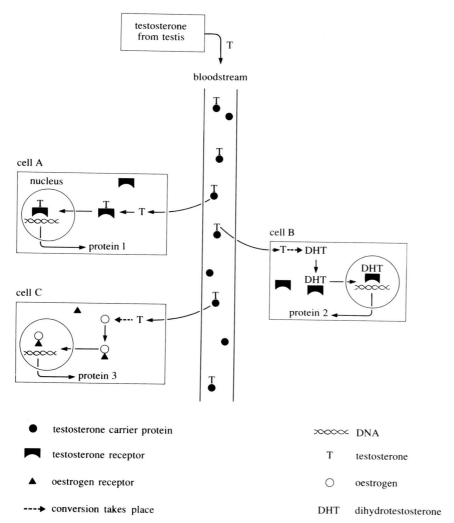

Figure 4.18 Diagrammatic summary of the action and conversions of testosterone. Testosterone (T) is mainly synthesised in the testes and secreted into the blood, where it is picked up by testosterone carrier protein. Testosterone is able to enter cells, where it acts on the cell's DNA, via the testosterone receptors, to produce protein 1 (cell A), or, after conversion to dihydrotestosterone (DHT), to produce protein 2 (cell B). Alternatively, testosterone is converted to oestrogen, when it acts on the DNA via the oestrogen receptors to produce protein 3 (cell C).

Throughout this discussion, the rather clumsy terms 'male-like' and 'female-like' have been used; there is good reason for this. You will recall that in Book 1, Section 9.3.1 it was stressed that it is very important not to assume that differences in reproductive behaviour between the sexes are consistent between animals. The stereotype that males are aggressive and sexually assertive, whereas females are docile and primarily concerned with parental care is highly misleading. An overview across a wide range of animals reveals that the various aspects of reproductive behaviour, such as courtship and parental care, and also including aggressive behaviour, are observed in both sexes. Within any one species, they may occur more in one sex than the other, to a greater or lesser extent. It is thus quite wrong to regard aggression, for example, as 'male behaviour', and, as this section has shown, it is equally wrong to regard testosterone as an exclusively 'male' hormone. The term 'male-like' means 'typical of males in the particular species being considered', such as the rat; it does not mean 'typically male' in a general sense.

Given that there are behavioural differences between males and females, it is not unreasonable to expect, and hence to look for, neural and, in particular, brain differences between males and females. Although a number of differences have been found, only three will be considered here, because they have been studied from a developmental perspective: the medial preoptic area of the hypothalamus, the amygdala and the spinal nucleus bulbocavernosus.

The medial preoptic area

Roger Gorski and his colleagues in California have studied the medial preoptic area of the hypothalamus and the effect of testosterone on that area early in life. There is a high density of neuronal cell bodies in the medial preoptic area in both male and female pre-natal rats. By about the first or second day after birth, the volume of the high-density area in the medial preoptic area is larger in male rats than in female rats, a difference that persists into adulthood. The density of neurons within the high-density area is the same in both males and females; it is the *volume* of that high-density area that is different. Gorski called this high-density area the *sexually dimorphic nucleus of the preoptic area*, SDN-POA. How does this difference in the SDN-POA arise? Gorski *et al.* (1978) took the classic approach of giving a single injection of testosterone to post-natal female rats and castrating post-natal male rats. Both treatments affected the volume of the SDN-POA (see Figure 4.19, treatment groups E and C, respectively). In the male, castration caused a 50% reduction in volume at 30 days old compared with normal males, a reduction that could be largely prevented by an injection of testosterone the day after castration (Figure 4.19, treatment group D). Thus, 50% of SDN-POA volume in the male is dependent on the presence of post-natal testosterone. The injection of testosterone into post-natal females (Figure 4.19, treatment group E) increased the SDN-POA volume considerably.

☐ What is the significance of the oil injection in treatment group C (Figure 4.19)? (*Hint* Testosterone was dissolved in oil before it was injected.)

■ Treatment group C is a control for treatment group D. The procedure used for the two groups was identical, except that in treatment group C only the inert oil was injected, whereas for treatment group D the oil contained testosterone.

The results from these experiments suggest that the volume of the SDN-POA is affected by testosterone, but some of the differences still need to be accounted for.

☐ What differences still need to be accounted for?

■ The volume of the SDN-POA in castrated males is greater than that of normal females (Figure 4.19, treatment group C compared with group A), and its volume in females treated with testosterone is not as great as that of normal males (treatment group E compared with group B).

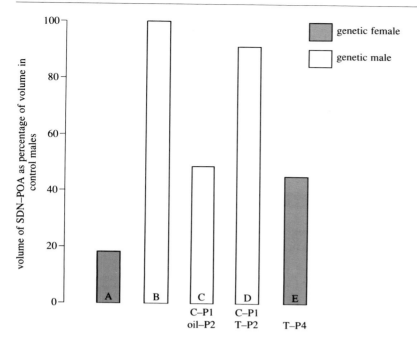

Figure 4.19   Summary of the sex difference in SDN-POA volume and the influence of the administration of post-natal testosterone. SDN-POA volume is expressed as a percentage of the volume of this nucleus in control male rats, 30 days after the treatment. Normal female rats are shown in A and normal males in B. In C, males were castrated on post-natal day 1 (C–P1) and given an injection of oil on post-natal day 2 (oil–P2); in D, similar males (C–P1) were then injected with testosterone on day 2 (T–P2). In E, females were injected with testosterone on post-natal day 4 (T–P4).

In the experiments cited so far the level of testosterone was altered some time after birth. It is still possible that testosterone is the key hormone in the volume differences in the SDN-POA between males and females, if, for instance, testosterone begins to exert its effects before birth. A more prolonged exposure to testosterone, beginning pre-natally, results in female rats with an SDN-POA volume equivalent to that of males. Thus, the differences in SDN-POA volume between male and female rats appear to be due to differences in the level of testosterone both pre-natally and post-natally.

The amygdala

The amygdala (Book 2, Section 10.4) is the second area showing neural sex differences. Michael Meaney in Canada has studied the neural basis of the sex differences in play-fighting in rats discussed above. He found that lesions to the amygdala on P21 or P22 reduce the amount of play-fighting in male rats to the level of that seen in females. The same lesions have no effect on the level of play-fighting in females. This finding suggests that a sex difference in the amygdala might mediate the sex difference in play-fighting. Two further pieces of evidence support this hypothesis.

Firstly, females given an injection of testosterone directly into the amygdala between P1 and P5 engage in slightly more play-fighting than normal males. Oestrogen does not have this effect, so testosterone itself exerts the effect.

Secondly, examination of the concentration of testosterone bound to its receptor in different parts of the neonatal brains of male and female rats reveals that the greatest difference in concentration between males and females is in the amygdala.

☐ Does this last piece of information accord with the finding that male rats with the Tfm mutation show female-like levels of play-fighting? (If you are uncertain about the Tfm mutation, refer to Section 4.2.)

■ Yes it does. The Tfm mutants have normal levels of testosterone but no testosterone receptors. In the absence of the receptors, the testosterone could not exert its effect on the amygdala.

The spinal nucleus bulbocavernosus

A third area showing neural sex differences is the spinal nucleus bulbocavernosus (SNB). The SNB is in the spinal cord rather than the brain, but it too is sexually dimorphic. Axons from this nucleus innervate the muscles involved in the penile reflexes of copulation. As you might expect, the nucleus is present in male rats and absent in females. So, how does this sexual dimorphism arise?

Arthur Arnold and Marc Breedlove at the University of California have studied this problem and arrived at a novel hypothesis. Both male and female rats have the appropriate muscles for the penile reflexes of copulation, the LA/BC muscles, when they are born. However, in females the muscles degenerate unless the neonate is supplied with testosterone or dihydrotestosterone. A similar picture emerges for the SNB nucleus; this too is present in both sexes at birth, but the neurons of the SNB nucleus in the female die off, whereas those of the male remain. Given the two previous examples, the SDN-POA and the amygdala, it would be reasonable to suggest that testosterone exerts its effect by sustaining (that is, promoting the survival of) the neurons of the SNB nucleus and sustaining the LA/BC muscles.

☐ There are two other related hypotheses as to how testosterone exerts its effect on this sexually dimorphic system. What are they?

■ One possibility is that testosterone sustains the neurons of the SNB nucleus and that those neurons are then able to maintain the LA/BC muscles. The other possibility is that testosterone sustains the muscle and the muscle maintains the neurons of the SNB nucleus.

In 1986, Breedlove presented two pieces of evidence in support of one of these hypotheses. Firstly, he produced mosaic rats in which the genotype of some of their SNB neurons contained the Tfm allele. These Tfm neurons do not produce the testosterone receptor, and so cannot respond to testosterone. Yet some of the Tfm neurons survived into adulthood. Secondly, in normal rats, testosterone receptors are not present in SNB neurons at the time the hormone prevents these neurons from dying.

☐ Which of the three hypotheses do these data support?

■ They support the hypothesis that testosterone sustains the muscle and the muscle then maintains the neurons of the SNB nucleus.

This example again illustrates the importance of the target in stabilizing the neuronal population and thereby matching the number of neurons to the size of the target (Section 3.2).

The above studies all illustrate the powerful effect of testosterone and oestrogen on the development of brain and behaviour. That is what the studies were selected to do. They have the virtue of having been replicated on numerous occasions, for which reason they are regarded as robust findings. However, it should be remembered that both testosterone and oestrogen have other effects besides those reported here, and not all of their effects are on the brain and behaviour. Furthermore, they are not the only hormones that have important developmental effects.

Before leaving this section on sex differences, it is worth stepping back from the detail to look at the broader perspective. The presence or absence of testosterone early in life clearly does have an effect on sexual behaviour and play-fighting in rats and monkeys. Furthermore, testosterone affects the volume of the SDN-POA and the functioning of the amygdala. What has not been demonstrated is how a change in the volume of the SDN-POA, for example, could affect behaviour, or how testosterone, by affecting the amygdala, affects play-fighting. What is the link between these brain changes and behaviour? At present, very little can be said about the link, but two things are clear. Firstly, neither of these brain changes directly alters the sensory pathways to the brain. Secondly, neither of these brain changes directly alters the motor pathways from the brain.

☐ What, then, do the changes alter?

■ The changes must alter the way information is processed within the brain.

To answer the question about the link between brain processing and behaviour means establishing the neural correlates of the behaviour, and this means discovering not just which nuclei are involved and what the connections are between them, but also understanding how the nuclei interpret the vast number of action potentials arriving every second. You will recall from Book 2, Chapter 7, that unravelling even the relatively simple neural network of flight in the locust is a major undertaking. Think how much more so this would be for a complex behaviour like play-fighting. Establishing how observed differences in the brain cause differences in behaviour is a major, as yet unsolved, challenge of neurobiology.

The final example in this section returns to a type of behaviour introduced in earlier books, namely bird song. This time, the development of bird song is considered from a neural perspective.

4.4.3 Bird song

You learnt about the behavioural development of bird song in Book 1, Section 5.6. Not all species of bird sing, but in those that do, singing is usually a sexually dimorphic behaviour: generally adult males sing a lot, adult females do not sing at all. Investigations of the neurological basis of this behavioural difference have used both lesioning and microscopic techniques. Lesioning studies have implicated two areas of the bird brain as being particularly concerned with song (Figure 4.20, *overleaf*). Removal of either the higher vocal centre (HVc) or the robustus archistriatalis (RA) area impairs singing in males but appears to have no effect on the behaviour of females. (These particular areas of the bird brain have no mammalian equivalent.) Microscopic examination of serial sections of brain tissue reveals a striking difference between male and female brains in both the HVc and the RA nuclei. In the adult male these areas are significantly (two or three times) larger than in the adult female. Thus, the behavioural difference would appear to have a neurological correlate.

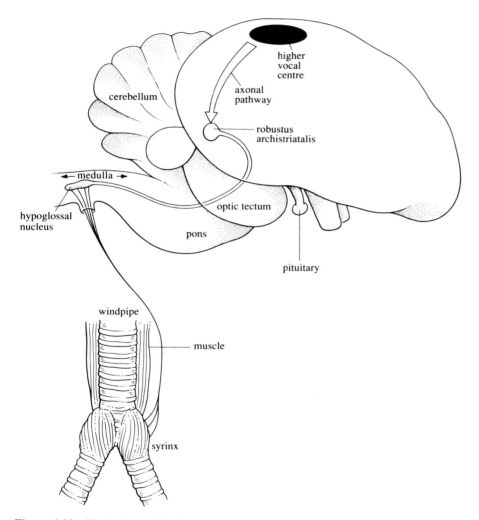

Figure 4.20 The brain nuclei of the male song bird which control singing. The largest is the higher vocal centre (HVc), which sends many axons to the robustus archistriatalis (RA). Axons from the RA in turn contact motor neurons in the hypoglossal nucleus, which innervate the muscles of the syrinx, the organ where the sound is actually produced.

How do these neurological differences arise, and what factors influence their development? A number of pieces of evidence have a bearing on this question.

In the zebra finch (*Taeniopygia guttata*) the adult female does not sing, and her HVc and RA nuclei are virtually absent. This extreme sexual dimorphism in the brain is of considerable benefit when looking at development, because the differences between male and female are conspicuous. Masakazu Konishi in California has found that at P5 there are no differences between male and female zebra finches in either the HVc or the RA nuclei. By P12 there is a noticeable decrease in the volume of the HVc area in females and by about P30 there is a decrease in the volume of the RA region in females.

A second piece of evidence is that testosterone or oestrogen given to female zebra finches during the nestling period (4–12 days after hatching, that is, days P4–P12), followed by a further injection of testosterone as adults, allows them to sing; an injection of testosterone into a normal adult female zebra finch does not lead to her singing. Thus, both testosterone and oestrogen can exert an organizing effect on the female zebra finch's brain during the nestling period. Testosterone also has an activating role in adulthood. Interestingly, castration of the young male has little effect, for when later injected with testosterone as an adult he will still sing.

☐   What does this last result imply?

■   That the source of the hormone that organizes the young male brain for future singing is not the testes.

☐   Why would this lead you to suppose that the organizing hormone in male zebra finches is oestrogen rather than testosterone?

■   Because the testes are the principal source of testosterone and yet the male zebra finch's brain is organized in the absence of the testes.

Support for this conclusion comes from an examination of the amount of these two hormones found in the blood of young zebra finches. Males and females contain the same, low amount of testosterone, but males contain considerably more oestrogen. The source of this oestrogen has yet to be determined.

☐   Given that oestrogen is the organizing hormone, would you expect dihydrotestosterone to have an organizing effect?

■   No, because dihydrotestosterone, unlike testosterone, cannot be converted into oestrogen in the cell (Figure 4.18). So dihydrotestosterone cannot mimic the effects of oestrogen. Dihydrotestosterone does not have an organizing effect on song.

If neonatal oestrogen administration alters singing behaviour in females, and the HVc and RA nuclei are absent in normal females, yet are important for singing, the next question to ask is: does neonatal oestrogen administration affect the size of the HVc and/or RA nuclei?

The answer to this question is, 'yes it does', and it does so in proportion to the timing and duration of oestrogen administration; the earlier it is done and the longer it is done, the greater the effect as measured by volume changes in the HVc and RA nuclei. If a brain nucleus increases in size, it can be the result of either or both of two things. There could be an increase in the *number* of neurons in the nucleus, or there could be an increase in the *size* of the neurons already present. Both of these processes appear to occur in response to the oestrogen administration. Indeed the HVc of the juvenile male (P40) appears to be still producing neurons when other areas of the brain are not.

In summary, it would appear that, as nestlings, male zebra finches have higher levels of circulating oestrogen in their blood than females. The oestrogen promotes the growth of the HVc and RA nuclei. In adult males these nuclei can respond to testosterone, and thus the males sing. Note that oestrogen here is playing the organizing role that testosterone plays in young rats and monkeys.

However, another conclusion is equally tenable, and that has to do with the fact that the syrinx (Figure 4.20) the muscular organ in the throat of the bird which produces the song, is also sexually dimorphic. In the male the syrinx is much larger and more muscular than in the female. The period when male zebra finches start practising song is approximately P35, well before the end of the learning phase, and overlapping with the increase in neuron number in the HVc of juvenile males. Furthermore, over 50% of the neurons that male zebra finches add to the HVc during song learning are projection neurons, which form part of the motor pathway for song production (the axonal pathway indicated in Figure 4.20). So could it be that here is another example, like the SNB nucleus, where the target, in this case the muscles of the syrinx, sustains the innervating neurons? Does a large syrinx result in a large higher vocal centre?

The following two sets of evidence should convince you either that oestrogen exerts its effects directly on the HVc or that oestrogen promotes the growth of the muscles of the syrinx, which in turn sustains the HVc. Read the evidence and decide which conclusion is correct.

1   Dihydrotestosterone administered to normal nestling males increases the size of their HVc by 50%. Dihydrotestosterone administered to normal nestling females does not alter their propensity to sing as adults. Dihydrotestosterone promotes muscle growth in the syrinx, but oestrogen does not.

2   In the swamp sparrow (*Melospiza georgiana*), unlike the zebra finch, the periods of song memorization and song rehearsal do not overlap. Neuron number in the HVc nucleus is plotted against age of bird in Figure 4.21. The sensitive period for song memorization in the swamp sparrow is up to P120. Rehearsal does not begin until about P275. Thus, the period of greatest increase in neuron number coincides with the memorization phase, not the motor phase when rehearsal or practice singing takes place.

☐   Does oestrogen exert its effects directly on the HVc or does oestrogen promote the growth of the muscles of the syrinx, which in turn sustains the HVc?

■   The crucial piece of evidence is that oestrogen does not promote muscle growth. So oestrogen could not promote growth of the muscles of the syrinx, with consequent effects on the HVc. Furthermore, even when a hormone that does promote muscle growth was injected into females (that is, dihydro-testosterone), there was no effect on their singing. So these females did not have the appropriate neurons and neuronal interconnections that would enable them to sing, despite presumably having a large syrinx.

If oestrogen were to exert its effect on the HVc through the muscles of the syrinx, then the increase in size of the HVc should coincide with the motor phase of song

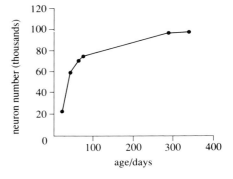

Figure 4.21   Plot of neuron number against age in the HVc nucleus of the swamp sparrow.

learning. However, the second piece of evidence shows that this is not the case: the increase in the size of the HVc coincides with the memorization phase.

☐   What is the significance of the 50% increase in the size of the HVc of normal nestling males following administration of dihydrotestosterone?

■   Dihydrotestosterone can affect the muscles of the syrinx. So it is possible for the muscles of the syrinx to affect the size of the HVc, as set out in the hypothesis above, but only in normal nestling males and with oestrogen also affecting the HVc.

Finally, Fernando Nottebohm in New York has studied the annual volume changes in the HVc and RA nuclei of canaries (*Serinus canaria*). Unlike zebra finches and swamp sparrows, which do not show these annual volume changes, male canaries amend and expand their repertoire of songs year by year. The volumes of the HVc and RA nuclei vary with the amount of singing the canary is doing, being greatest in the spring and least in the winter. There are no volume changes in the HVc and RA nuclei of the female canary.

What Nottebohm found, using the technique of injecting radioactive thymidine (Section 2.3.4), is that new neurons are born in adult male canaries. In fact, the process exactly matches that seen in young birds (compare the process described for young mice in Chapter 2), where cell division occurs at the ventricular zone and the new neurons migrate to the HVc, but not, apparently, to the RA nucleus. The brain of the adult canary does not increase in size year by year, which means that an equivalent number of neurons to those born must be lost each year. The finding that new neurons are born in the brain of the adult male canary contradicts the orthodox view that new neurons are not born in maturity. The finding also raises some interesting questions. One is how the canary achieves this feat. Another is how prevalent is the phenomenon. Answers to these questions await further study.

This section has focused on the production of song. However, the song can only function in communication if there is also a receiver to hear and interpret song (Book 3, Section 2.2). These tasks presumably involve certain other brain regions, but as yet, little is known about them.

Summary of Section 4.4

Hormones can affect the growth and connections of neurons made during development. In moths, juvenile hormone can inhibit the growth of sensory axons. In mammals, testosterone affects the size of the sexually dimorphic nucleus of the medial preoptic area and oestrogen has a similar effect on the higher vocal centre in birds. Testosterone also appears to organize the amygdala in rats. It also appears to affect the spinal nucleus bulbocavernosus of the rat, but by its effect on the muscles involved in the penile reflex rather than by any direct effect on neurons.

This catalogue of effects of testosterone and oestrogen on neural structures attests to the importance of these hormones during development. These neural effects are paralleled by effects on behaviour patterns as diverse as play-fighting, sexual behaviour and singing in birds. It is not yet clear how differences in the size of a brain nucleus, or its internal connections, result in differences in behaviour. Filling in all the gaps in the developmental jigsaw between the hormone, its neural effects and the eventual behaviour of the adult remains a daunting challenge for neuroscientists.

4.5 Human development

In this final section of Chapter 4 the subject of human development is briefly addressed. Perhaps the most striking thing about human development is that despite its complexity, it is completed normally most of the time. Unfortunately, there are occasions when development does not proceed normally, and reasons are then sought as to why development went wrong. The study of human development differs in one major respect from the studies considered so far in this chapter: human development is not interfered with experimentally. Thus, the study of human development relies on the careful observation of differences between individuals that arise naturally. In essence, this often means that abnormal development is studied to provide clues to its prevention, and hence to an understanding of the normal developmental process. Sometimes particular causal agents of abnormal development can be identified (for example certain drugs, or diseases) from the vast number of possible agents, and sometimes not. Focusing on the problems of development, as this section does, tends to give a distorted picture of the incidence of these problems. Most of those considered here are relatively rare events and, as this section is considering the developmental process, statistics relating to incidence are not given.

The section begins with a very brief description of normal motor development up to about two years. The rest of the section is then roughly divided to match the previous sections in this chapter. Section 4.5.2 looks at some genetic influences on development; Sections 4.5.3 and 4.5.4 consider external influences on development; Section 4.5.5 returns to the effect of testosterone on the development of sex differences (see also Book 6, Chapter 4).

4.5.1 The normal course of development

For each individual the advent of a particular ability is an exciting and important event. When a particular ability will appear cannot be predicted, and the developmental history of each individual is therefore unique. Putting together the developmental histories of many individuals does, however, allow average patterns of development to emerge. These patterns can then be used to assess whether an individual developmental history fits the average pattern, or does not fit the pattern. The word normal is often used to mean 'fits the average pattern' and the word abnormal to mean 'does not fit the average pattern'. Abnormal and normal will be used in this sense in this chapter.

Development clearly begins pre-natally, and Figure 4.22 shows some of the key events during the period of gestation. Note in particular that the fetus can respond to stimulation at 10 weeks, that the sex organs are distinct by 16 weeks, and that major body movements occur from 22 weeks. (Note that the figure is for reference only; you do not need to remember the details.)

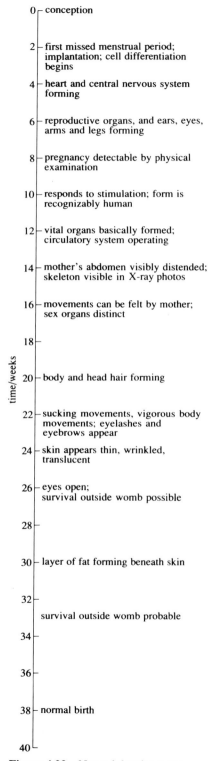

Figure 4.22 Normal development: some pre-natal milestones.

It is worth reiterating that, in biological terms, immature individuals, including children, are not merely small adults. They are individuals who have to survive with the abilities they have at a particular age. They are also growing into adults. These two activities, surviving and growing, are quite distinct.

To take obvious examples, at birth a child has a rooting reflex; that is, it will spontaneously suck at an object of appropriate size and shape, like a nipple or a finger. Later in development, the child will not suck, but chew. But sucking involves quite different sets of nerves and muscle movements from chewing. A sucking baby is not practising sucking in order to learn how to chew, but is sucking in order to exist as a baby. Similarly a child learns to crawl before it walks, but crawling is not just practice for walking. On the other hand, as a child begins to speak, it is clearly rehearsing and practising communication skills that will increase with age. So some skills are for the here-and-now, and some skills are for the future.

With this caveat in mind, the milestones by which developing children pass are now described. A new-born baby lies horizontally unless held. During its first year of life it learns to sit upright, and during the next years it acquires different physical skills, such as running, stopping, climbing, jumping and so on. A new-born baby can only open and close its hand, but by the end of the first year of life it can pick up tiny objects neatly, transfer them from hand to hand, mould its hands around them, poke its index finger into parts of interest and explore the objects with its hands. During the next few years the child acquires the ability to hold objects more precisely and to use its hands skilfully. For example, a baby of 4–5 months will try to pick up an object by swiping at it with both hands, often transmitting its excitement to its whole body. Contrast this with the efficient way a 5–6-year-old can use a pencil to draw a very small circle.

How far can a child's acquisition of these skills be related to the development of the brain? Neurons in different parts of the brain mature and make connections with one another, and with the sensory inputs and motor outputs at different times during development. In particular, there is increasing control over body processes by the cerebral cortex, so that the reflexes (like the grasp reflex) that the month-old baby shows disappear, to be replaced by more conscious, coordinated holding of objects. Even a child only a few days old can turn its head towards faces and sounds. By three months it can focus its eyes on an object and move its hands towards it, and can be soothed by a familiar voice. During this time the regions of the brain responsible for coordination of hand and eye, and for accurate recognition of sounds, are beginning to develop.

The sequence in which the normal child acquires successive motor skills is fairly predictable, and has led to the concept of development proceeding in stages. This concept is somewhat misleading, because development is a *continuous* process. (There is, after all, no identifiable point at which one stage finishes and another begins.) Furthermore, there are many developmental paths to maturity, so, for reasons given in Book 1, Chapter 5, you should not think of development as proceeding in an inevitable way along a prescribed path. The sketches shown in Figure 4.23 (*overleaf*) present a few stills from the 'film' that represents the developmental sequence. If you have children of your own, or have watched other people's children, you will realize that the ages mentioned are only approximations.

Figure 4.23 Normal development: some post-natal milestones. Note that (a) the process is one of continuous development, not steps and stages; (b) the sequence is the same for all children (for example, head control always comes before sitting, sitting before crawling, etc.), though there are many individual variations (for example, some babies do not crawl but progress straight to walking); (c) the age at which a particular motor skill is achieved varies within certain limits.

There are, of course, many changes occurring inside the body which are not reflected in these outward signs. Of particular interest are the hormonal changes, which are considered in Section 4.5.5.

4.5.2 Some genetic influences on human development

In Book 1, Section 3.3.3 you were introduced to the disease phenylketonuria (PKU). The disease results when people are unable to make the enzyme phenylalanine hydroxylase because they are homozygous for a particular pair of recessive mutant alleles. As a result, an inactive form of phenylalanine hydroxy-lase is produced. PKU is a genetic disease, sometimes called an *inborn error of metabolism.* Numerous inborn errors of metabolism are known, but only PKU and Tay-Sachs disease will be considered here. (Others are considered in Book 6, Chapter 4.)

Before they are described further, though, the genetic diseases must be distin-guished from another class of diseases, the **chromosomal diseases**. The distinction between genetic diseases (that is, where a genetic mutation is inherited) and diseases resulting from abnormal chromosomes is necessary because a mutation is a small change in one gene, whereas the addition or loss of a chromosome represents not only a great number of genes, but also a complication to the normal cell division mechanism. (Incidentally, both genetic and chromosomal diseases may be congenital. **Congenital diseases** are those for which the symptoms are already present at the time of birth. Whatever the cause of these diseases, be it genetic, too many or too few chromosomes, or errors in fetal development, the symptoms are present at birth.)

Two examples of diseases resulting from chromosomal abnormalities are trisomy 21 and Turner's syndrome.

Trisomy 21 is better known as Down's syndrome. Children with Down's syn-drome are happy, loving and friendly. They also have retarded mental and motor development, broad flat faces and a noticeable fold along the inner corner of the eye. Instead of having a *pair* of a particular chromosome (chromosome 21) they have *three*.

Turner's syndrome, shown by women with only one sex chromosome, is characterized by their being short, with a broad neck and lacking all or most of the female reproductive organs and secondary sexual characteristics such as breasts. Normally a human sperm carries 23 chromosomes (one of which is a sex chromo-some, either an X or a Y). Occasionally the sperm that fertilizes the egg contains only 22 chromosomes, none of which are sex chromosome. The resulting zygote therefore only contains 45 chromosomes (22 from the sperm and 23 from the egg) instead of the normal 46. Such a zygote has only one sex chromosome, the X chromosome from the egg. (The normal female has two X chromosomes, and is designated XX.) The single-X female (designated X0) then develops Turner's syndrome.

It remains unclear how the chromosomal abnormalities in trisomy 21 and Turner's syndrome exert their effects on the developmental process. The two genetic diseases will now be considered.

Phenylketonuria

If phenylketonuria is untreated, it can result in reduced brain development and irreversible mental retardation (Book 1, Section 3.3.3).

☐   How can the effects of PKU be prevented?

■   By detecting PKU early in life and eliminating phenylalanine from the diet (in particular, by avoiding meat).

A successful treatment for PKU has existed long enough for there now to be individuals with PKU who are sexually mature and quite normal. The mature central nervous system, in contrast to the developing nervous system, appears quite resistant to the toxic effects of a high level of phenylalanine, so adults with PKU do not need to follow a strict phenylalanine-free diet. However, a fetus growing inside a PKU female may be born with severe mental retardation even though the fetus has normal phenylalanine hydroxylase, inherited from the father. If the mother is not on a low phenylalanine diet, the high concentration of phenylalanine swamps the fetus and causes abnormal neurological development. The child is born with mental retardation resulting from PKU, even though the child itself does not have the genetic disease.

Tay-Sachs disease

Tay-Sachs disease is a genetic disorder that results in death by about two years of age. Babies with Tay-Sachs disease appear normal at birth and usually continue to look healthy and normal until about six months of age. The startle reflex, whereby a baby extends its arms in response to a loud noise, disappears in normal babies by about four months of age; in babies with Tay-Sachs disease the reflex becomes more exaggerated and is initiated by very gentle sounds. Motor development may proceed normally, with the infant gaining head control and being able to sit unsupported by six months, but both of these abilities are lost soon after the first birthday.

The recessive gene involved in Tay-Sachs disease codes for the enzyme hexosaminidase. In affected individuals the mutant alleles they carry code for a non-functional version of the enzyme. The absence of this single enzyme results in changes in the nervous system. Neurons accumulate a particular kind of fat to such an extent that they become swollen, and this interferes with their function. As yet, there is no cure for Tay-Sachs disease.

The two examples presented here are very different from those presented in Section 4.2. They illustrate the biochemistry of the genetic defects, whereas the examples in Section 4.2 illustrated anatomical differences in the nervous system (for example in the macroglomerular complex of the sphinx moth). All the examples, though, demonstrate the influence of the genotype on development.

4.5.3 The pre-natal environment

The developing fetus is affected by the state of wellbeing of the mother. Any substance that can pass from the mother's bloodstream through the placenta (the organ by which the fetus is attached to the mother) and into the bloodstream of the fetus could affect the fetus. Whatever a pregnant woman eats or drinks, whatever drugs she takes, whatever she breathes in, may find its way to the fetus. Her state of nourishment, general health, level of anxiety or exposure to harmful substances may affect the development of her baby.

Despite these potential threats, the fetus is usually sufficiently well protected and buffered by the mother and the placenta to escape any noticeable effects, and thus develops normally. Only when conditions are extreme or in adverse combination is development conspicuously affected.

The extent to which the development of the fetus can be affected by a particular factor is called its *vulnerability*. Vulnerability is dependent on three things:

1 The fetus. Fetuses differ one from another in the same way as any individuals do. Thus, the response of a particular fetus to a drug, say, may differ from that of another fetus exposed to the same drug.

2 The developmental stage at which exposure to the agent occurs. Vulnerability to a particular factor changes with the stage of development of the fetus. The vulnerability of the fetus to alcohol, for instance, is more pronounced later in development than earlier. One crucial point here has to do with when particular organs or systems in the fetus are developing. This is shown diagrammatically in Figure 4.24. Note, in particular, the times when the ear and the external genitals are growing, since these are referred to later. It is generally thought that a structure is most vulnerable during the period of its most rapid growth.

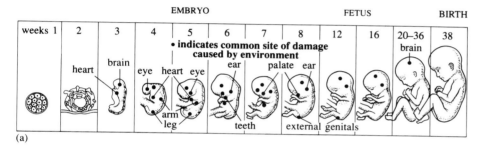
(a)

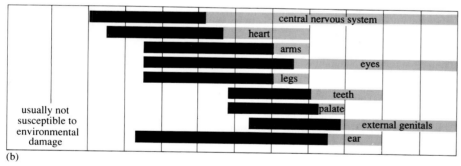
(b)

Figure 4.24 Fetal development of particular organs during development (a) and their vulnerability to environmental factors (b). The dark horizontal bars represent the period when the organ is being formed and is therefore most vulnerable. Organs are still susceptible in the period indicated by the light bars.

3 The amount of exposure. There are two components to exposure. The first is the quantity or concentration of the agent in particular organs of the body (or extent of absence if the problem is a deficit). Generally, the higher the level of exposure, the greater the effect. The second component is the duration of exposure. The longer the duration of exposure to alcohol, for example, the greater its effects. Remember that what may be a small dose for the mother, may be a large dose for the fetus.

Before going on to some of the factors that do cause alterations to development, reflect for a moment on the resilience of the fetus.

The ability of the fetus to withstand adverse conditions is well illustrated by the heroic study by a group of Americans led by Zena Stein. Stein *et al.* (1972) examined the extensive records of the Dutch authorities on births, birth dates and birth places. They were also given access, by the Dutch Defence Department, to records of tests carried out on male 18-year-olds as part of the military induction process. In all, the study population was 125 000 males born between 1 January 1944 and 31 December 1946. Those who survived and were still resident in the Netherlands when they were 18 were tested by the military (hence the male bias in the sample); some 98% of those who could be tested were tested. The main point of this study, and its relevance here, is that certain cities in the Netherlands under German occupation experienced an acute famine from October 1944 to May 1945, the *hongerwinter*. During the famine, food rations were down to 1 880 kilojoules per person per day, roughly one quarter of the recommended daily energy intake for an adult.

Although the number of births was severely reduced because of the famine, there were nevertheless several thousand of them. Stein *et al.*, however, could find no effect of the famine on intellectual performance at age 18, the measure of behaviour they looked at. Conception and development took place at various times during the famine; children were born and raised during and after the famine, yet there were no differences in the ability to perform the various tests of mental ability later in life. The Stein study focused on mental ability, so it is not possible to say that there were no other effects on development because of the famine. The overwhelming impression from this study, though, is that the 18-year-olds were normal and no different from those born in parts of the country not affected by the famine. Thus, many developing fetuses had been resilient enough to develop normally despite their mothers' privations.

☐   The famine occurred during the Second World War and so cannot be taken as evidence that nutrition has no effect on mental performance. Why not?

■   It is possible that *all* those born in the Netherlands during the period 1944–6 were poorly fed and under stress. Thus, the control group might have under-performed compared, say, to a group well fed during this period.

Support for this idea comes from a study in which American children conceived in the autumn and winter were found to be heavier, healthier and more likely to go to college and appear in *Who's Who in America* than children conceived in the spring and summer. Diet was considered to be the main causal factor here, with the mothers eating protein-rich stews and roasts in the winter, whereas they tended to eat salads and fruit in the summer. This cannot be regarded as hard evidence one way or the other (after all, temperature also varies from summer to winter), but it is suggestive, and highlights the problems of trying to ascribe cause and effect in human development.

These studies have concentrated on nutritional quantity, but quality is also important. Some vitamins and all minerals have to be obtained from the diet because they cannot be made in the body. One familiar example is vitamin C, which is an essential dietary ingredient for humans and other primates. A less familiar example is folic acid, the lack of which has recently (1991) been implicated in certain developmental disorders called neural tube defects (for example *spina bifida*, a congenital disease in which part of the spinal cord is exposed through the back). Taking folic acid during pregnancy does not guarantee that the fetus will not develop neural tube defects, however, though it significantly reduces the likelihood of such defects occurring. In many parts of the world, such as China, Spain, Ecuador and New Guinea, a common mineral deficiency is lack of iodine. In the Jimi Valley in New Guinea there was a marked increase in the number of children with *cretinism* during the 1950s and 1960s: the symptoms were severe mental retardation, stunted growth and deafness. The cause of the symptoms was confirmed as a lack of iodine in their diets, and, more importantly, a lack in the mothers' diets during pregnancy. The traditional salt used in the Jimi Valley was sea salt, which is rich in iodine. This salt, which was the only source of iodine in the children's diet, had been replaced by the more easily available refined Western salt, which lacks iodine. Without iodine the fetuses' thyroid glands did not develop normally. (The thyroid is an endocrine gland that secretes the hormone thyroxine.) Once iodine was given to adults in the affected areas, as iodized salt, there were no further developmental problems in their subsequent children. (Some of these studies are considered again in Book 6.)

Diet, then, clearly has an important influence on development. Any other external factor that interferes with development is called a **teratogen**. One example of a teratogen is the common viral disease rubella, or German measles, which results in a minor illness if caught by an adult or child, but it is serious if caught by a pregnant woman. The woman herself is only slightly debilitated by the disease, but the fetus can be profoundly affected. The effect of the virus on the fetus varies with the age when the fetus is infected, as shown in Figure 4.25. The authors of the study from which this figure was taken (Munro *et al.*, 1987) were careful to distinguish maternal symptoms of the disease (rashes) from fetal infection, since maternal symptoms may occur some time before fetal infection. Thus, the figure refers to the age at which the *fetus* was infected. Note also that rubella can cause multiple defects.

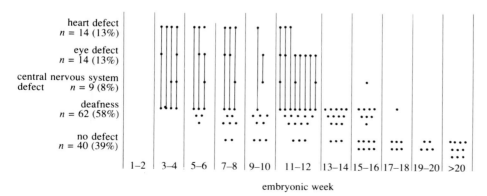

Figure 4.25 Effect of rubella infection on the fetus at different ages of development. The defects indicated were detected in infants who were known to have been infected with rubella as fetuses. Each dot represents an individual, except where dots are joined by lines, which indicates multiple defects in one individual. Data for 106 infants are presented. *n* = the total number of individuals with a particular defect (also shown as a percentage of the total number of infants in the sample).

☐ Between what ages does rubella infection affect development?

■ Rubella infection can alter development between fetal ages 3 weeks and 18 weeks.

☐ Does rubella infection during this time always result in defects?

■ No. Several fetuses infected between 7 weeks and 18 weeks were unaffected by the infection.

☐ Is the vulnerable period of the heart to rubella infection the same as the vulnerable period of the ear to rubella infection?

■ No. The heart is vulnerable between weeks 3 and 12, whereas the ear is vulnerable between weeks 3 and 18.

These clinical data show two things. Firstly, different individuals are susceptible to rubella infection to different extents. For example, some individuals were unaffected by an infection in weeks 7–8, whereas others suffered multiple defects. Secondly, different organs are susceptible to rubella infection (that is, are vulnerable) at different times during development. Exactly how the rubella virus affects development remains uncertain, though it has been suggested that it interferes with normal cell division.

The most infamous teratogen is thalidomide (Box 4.2). This drug was marketed in 1961 as a treatment for the sickness associated with the early months of pregnancy. Thus, the drug was specifically taken during the early stages of the development of the fetus. Somehow the drug prevented the normal development of the limbs and some 10 000 infants were born with abnormal limbs.

Box 4.2 Thalidomide

Thalidomide is now known to exist in two forms, which have the same structural formula but are mirror images of each other. The R-(+) form (Figure 4.26 left) has therapeutic value, but the S-(−) form (Figure 4.26 right) is toxic. The drug was supplied as a mixture of the two forms.

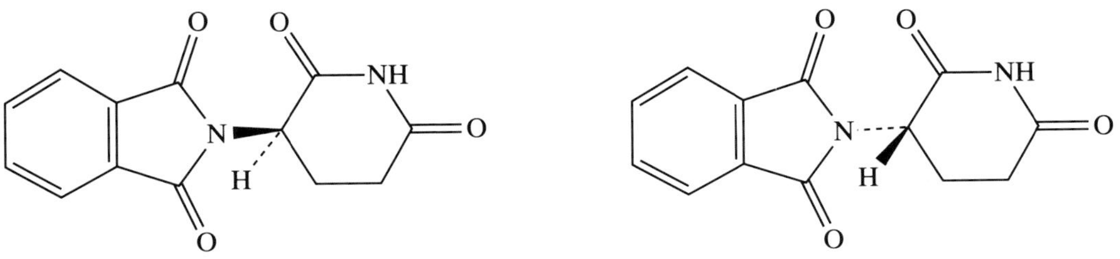

R-(+)-thalidomide S-(−)-thalidomide

Figure 4.26 The two forms of thalidomide.

Another drug, the steroid diethylstilbestrol, was routinely administered to pregnant women who were at risk of having a miscarriage. The drug worked, in the sense that it prevented miscarriage, but it also altered the development of the sexual organs. Some 60% of the daughters whose mothers had received diethylstilbestrol during pregnancy developed abnormal vaginal tissues.

A common teratogen is alcohol. Fetal alcohol syndrome was identified in 1973 among infants born to women who drank heavily during pregnancy. Infants with fetal alcohol syndrome are typically short, mentally retarded and slow in motor development. Exactly how alcohol exerts its effect on the developing fetus is not known, but it readily crosses the placenta into the bloodstream of the fetus. Once in the fetal bloodstream, it remains there for a long time, since the fetus's liver is unable to break down the alcohol. The impact of alcohol on the fetus can be gauged from the following study in which the effect of one drink of alcohol on the breathing movements of the fetus was studied. A group of women, all between 37 and 39 weeks pregnant, were given a drink. Half the women drank ginger ale, and the other half drank ginger ale to which one fluid ounce of 80% proof vodka (this is equivalent to one Scottish pub measure or slightly more than one English pub measure) had been added. In all the women who drank the vodka, their fetus stopped its breathing movements sometime between 3 and 30 minutes later. Many of the fetuses did not make breathing movements again for half an hour. As the level of alcohol in the mother's blood dropped, so normal breathing movements resumed.

☐   What was the significance of the breathing movements in this study?

■   The change in breathing movements was a measure of the effect of the alcohol on the fetus.

It is unlikely that this brief alteration in breathing movements was harmful to the fetus, or indeed that the cessation of breathing movements *per se* affected development. What is clear is that alcohol can affect the fetus.

Rather more is known about what substances are teratogenic than about the mode of action of a particular teratogen, though the effect of diethylstilbestrol is considered further in Section 4.5.5. You may be aware of the mismatch between the gross effects of teratogens given here, and the detail of neuronal growth and development given in Chapter 2. This mismatch results from the fact that at the time of writing (1992) there has been little study of the neuronal basis of teratogenic effects, and represents a real gap in the current understanding of human development.

4.5.4   Sensory development

Two senses are considered here as examples of the way sensory systems develop in infants: hearing and sight. To reach any sensible conclusions as to how any sensory system develops in human infants, three particular problems have to be overcome. The first problem is that infants are neither very mobile nor very responsive. If a particular stimulus does not elicit a response, then that might be because the infant did not detect the stimulus (in this context, an interesting result), or it might be because the infant was unable to respond, having poorly developed motor control. Both these possibilities assume, of course, that the stimulus, if detected, is sufficiently interesting to the infant to elicit a response. It is necessary to distinguish between these possibilities.

The second problem has to do with the reliability of the response. Put another way, on what percentage of presentations must a stimulus elicit a response for valid conclusions to be drawn about the infants' ability to detect that stimulus? Clearly, the task of the psychologist here is to compare different stimuli and note responses to those different stimuli, but this does not entirely overcome the rapid changes in motivation seen in infants. Finally, and particularly with very young infants, the effect of other factors on their performance must be considered. If their mother received an anaesthetic during childbirth, this can slow the responses of the infant for a few days. A major cause of stress, such as circumcision, can also alter performance for a considerable time afterwards. The task of studying the development of infants, then, is not easy, and relies not only on their voluntary compliance, but on that of their parents too.

The visual capacities of the newborn are not very great. The visual system itself, including the eyeball, is immature. There is no fovea, and the muscles controlling the lens are weak, so that the image on the retina is not sharply focused. Their best image is formed for objects 10–25 cm away; objects further away can be seen, albeit less clearly. Figure 4.27 gives you some idea of what a one-month-old infant can see.

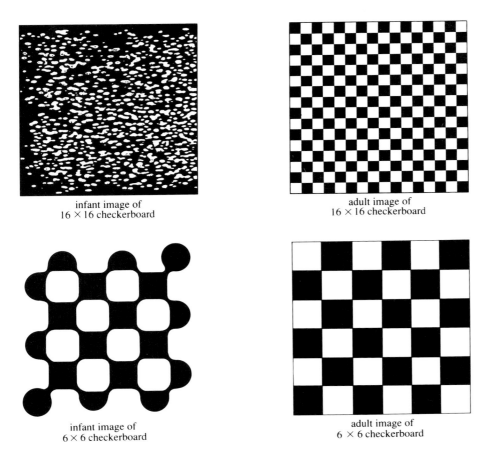

infant image of
16 × 16 checkerboard

adult image of
16 × 16 checkerboard

infant image of
6 × 6 checkerboard

adult image of
6 × 6 checkerboard

Figure 4.27 Response of the visual system of a month-old infant to two visual patterns compared to that of an adult.

The development of visual acuity (that is, the ability to see fine detail) has been studied by using the natural tendency of infants to suck, and to suck faster when the visual stimulation changes. By using a special type of pressure-sensitive artificial nipple, the rate at which an infant sucks can be measured. Tests were conducted on infants at various ages between two weeks and six months. While the infant is sucking on the nipple, two slides of the same size, brightness and colour are shown in quick succession to it, and the rate of sucking is then recorded. The slides have parallel lines on them. If the width of the lines on the second slide is the same as the width of the lines on the first slide, then there is no change in stimulation and no change in sucking rate. By changing the width of lines on one of the slides on successive presentations, the point at which the infants respond to the change by increased sucking marks the point at which they detect a change in stimulus, and hence is an indication of their visual acuity. When two weeks old, infants could distinguish between stripes that differed in width by 0.3 cm; at six months, they could distinguish stripes differing by 0.06 cm.

Does this change in visual acuity require visual experience or is it part of the natural maturation of the visual system? It appears that visual experience is not necessary for these changes, at least up to 12 weeks of age. Premature infants have visual abilities on a par with full-term infants of the same developmental age, despite having four or five weeks' additional visual experience.

Another question that can be asked is whether the human visual system has the same kind of sensitive period as that demonstrated in cats and monkeys (Section 4.3). There is evidence that abnormal visual experience during infancy and early childhood can result in permanent deficits in adult visual acuity (hence the importance of correcting squints in children referred to in Section 4.3.3). In particular, adults with a history of early visual deficits, such as astigmatism or myopia, exhibit lower acuity than normal adults even if the defect was corrected during childhood. So when is the sensitive period? It certainly lasts longer than a year and possibly as long as four or five years, but exactly how long is not known. Because the duration of the sensitive period is not known, visual deficits are corrected as early as possible to ensure that the correction takes place within the sensitive period, and to maximize the likelihood of normal adult vision.

The hearing of infants is very good, and in some respects remarkable. Infants will turn their heads towards a sound source within minutes of birth, suggesting that the ability to locate a sound source is present very early in life. Other aspects of hearing develop fairly rapidly. By two weeks, infants can discriminate the human voice from other sounds, and by four weeks they can discriminate between small differences in spoken sound. In a study by Peter Eimas and his colleagues in 1971, babies were trained to suck on a nipple to hear a recording of 'b' sounds. Eventually they habituated to this sound and sucked less. Gradually the recording changed to 'p' sounds. When the sound was clearly 'p' and not something between 'b' and 'p', the infants started sucking again. This study shows not only that these infants can discriminate 'b' from 'p' sounds, but also that the same sound categories that adults use are recognized.

These sound categories appear to be used by all infants and are not subject to aural experience. In cross-cultural studies, infants from many different linguistic backgrounds can discriminate between similar sounds. Some cultures use sounds that cannot be discerned by adult members of other cultures; yet infants of other cultures have no difficulty. For instance, Japanese adults find it difficult to discriminate between 'l' and 'r', but their infants can make the distinction easily. By one year of age, though, infants are becoming more like their parents, and have difficulty in distinguishing sounds to which they are not normally exposed. Many perceptual abilities improve with age, but some, it would seem, deteriorate.

4.5.5 Biology and gender

Chromosomes are the major biological indicator used to assess someone's sex. Chromosome tests are frequently used to test athletes in women's events to see if they 'really' are women. For the majority of people, chromosomal sex is unambiguous. Two of the 46 chromosomes that humans usually carry within their cells are described as sex chromosomes because, among other things, they contain the genetic information involved in creating either testes or ovaries. They therefore influence the kind of hormones produced by the ovaries or testes, and in turn the sexual characteristics of the person. Generally speaking, female body form results if the sex chromosomes are both X, and the male body form if the sex chromosomes are an X and a Y.

Consider those individuals who have two X chromosomes and one Y chromosome (that is, they have 47 chromosomes with an XXY complement of sex chromosomes), and other individuals who have just an X sex chromosome (that is, they have 45 chromosomes, with a sex chromosome complement of X0). XXY individuals look like males, whereas X0 individuals look like females, who will develop Turner's syndrome (Section 4.5.2).

☐ What do these observations suggest about the relative roles of the X and Y chromosomes in determining phenotypic sex?

■ A pair of X chromosomes does not necessarily confer femaleness, whereas the presence of a Y chromosome *does* confer maleness.

Researchers looking for genes that determine maleness have focused on the Y chromosome. A study of individuals carrying Y chromosomes in which different parts are missing has shown that an entire Y chromosome is not necessary for maleness. The apparently crucial, so-called 'male-determining' region is situated near its tip, and the gene here is thought to code for *testis-determining factor*. Sometimes it appears that this small region is transferred from the Y chromosome to an X chromosome. The result is an XX male, albeit one that can be explained. A certain proportion of XX men, and all true hermaphrodites (individuals with one ovary and one testis), do not, however, have the testis-determining factor.

These exceptions aside, the normal course of sexual differentiation is as follows. Both male and female embryos develop in exactly the same way until the seventh week of gestation. Each embryo develops two primordial gonads (primordial, because at this stage they are neither testes nor ovaries), and two sets of reproductive ducts (tubes), comprising the Wolffian duct, potentially male, and the Mullerian duct, potentially female. If the embryo contains the testis-determining factor, the primordial gonads become testes; in the absence of testis-determining factor, the primordial gonads become ovaries. Once the gonads have formed, one set of ducts develops while the other regresses. Which develops and which regresses depends on whether testes have been formed or not. Three hormones are secreted by the testis: testosterone causes further differentiation of the testes and the Wolffian duct; anti-Mullerian hormone demolishes the Mullerian duct; and dihydrotestosterone induces the masculine pathway of development of the external genitals. In the absence of testes, or more particularly, the hormones secreted by the testes, the Wolffian duct regresses, and the Mullerian duct and ovaries differentiate further.

An alternative model of how testis-determining factor works has been advanced by Ursula Mittwoch in London. She has suggested that, rather than acting as a switch, testis-determining factor is an accelerator; it accelerates the rate of growth of the primordial gonads. Thus, what determines whether a primordial gonad turns into a testis or an ovary is its size at seven weeks. If it is large, it becomes a testis; if it is small, it becomes an ovary. This attractive idea could certainly explain how hermaphrodites arise, but it is not generally accepted.

Testosterone has been prominent in this chapter both because its effects are profound and, as a consequence, because it has been well studied. However, before continuing to examine its effects, three points need to be made. Firstly, testosterone is not the *male* hormone. It is secreted by females, albeit in small amounts in most species, but in large amounts in others, and it is very often converted to oestrogen inside cells, with the oestrogen affecting the cell. Secondly, numerous other hormones that have not been considered here also influence development (for example thyroxine and growth hormone). Finally, development requires an orchestra of factors to act in concert. In focusing on the soloist (testosterone in this instance), one should not forget its dependence on the other players.

The role of testosterone in the development of behaviour was considered in Section 4.4, and in the formation of internal reproductive ducts above. Testosterone, after conversion to dihydrotestosterone, also has profound effects on the development of the external genitals. Figure 4.28 (*overleaf*) sketches the changes in the external genitals that occur in the human fetus as development proceeds. (You do not need to remember the details of this figure.)

Depending on how much dihydrotestosterone is present, the extent of development and the stage it has reached, the appearance of the genitals can vary between the male and female forms. Genital appearance at birth affects how an individual is brought up and how that individual behaves, as will be discussed below.

Briefly consider the following, which are actual cases from clinical records.

(a) A child was born with a congenital condition in which it had XY chromosomes and produced testosterone during development. However, the genitals and external appearance at birth were female. The child was therefore reared as a female. As a child she tended to prefer playing with dolls; as an adult she thought of herself unequivocally as a woman, even though she could not have her own children.

(b) A woman with a threatened miscarriage was prescribed a certain hormone (diethylstilbestrol). The (female) child was born with an enlarged clitoris that resembled a penis, so that there was some doubt at birth about her sex. She was however, raised as a female. As she grew older, she engaged in much more rough-and-tumble play (tomboyish play) compared with other girls from normal pregnancies.

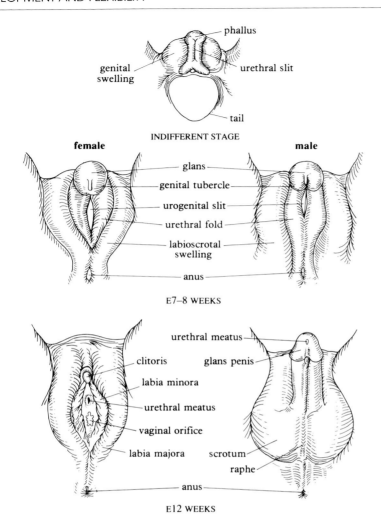

phallus

genital swelling

urethral slit

tail

INDIFFERENT STAGE

female   male

glans

genital tubercle

urogenital slit

urethral fold

labioscrotal swelling

anus

E7–8 WEEKS

urethral meatus

clitoris

glans penis

labia minora

urethral meatus

vaginal orifice

labia majora

scrotum

raphe

anus

E12 WEEKS

Figure 4.28 Changes in the structure of the external genitals in female and male human fetuses. The two sexes have clearly differentiated by the twelfth week after conception.

(c) In an isolated community in the Dominican Republic (Central America) there are people who change from female to male. That is, they are born appearing to be girls, but at puberty, they undergo all the bodily changes that are associated with male puberty, eventually looking and functioning like adult men. They subsequently seem to accept their role as men quite happily.

☐ Look carefully through this list of observations and write down whether you think that any of them unambiguously indicate that hormone exposure in the early life of a human being affects the brain in such a way that later gender-related behaviour is affected.

■ You might first have considered (b) and (c) as possible instances of early hormone exposure. In example (a), although testosterone was present, the child grew up believing herself to be female.

The particular anomaly of development cited in (a) is testicular feminization (Tfm): although there are testes present in the abdomen, which secrete normal amounts of testosterone, there are no receptors for testosterone, so the hormones do not act to push development in a masculine direction. (Note that both testosterone and dihydrotestosterone use the same receptor—called an *androgen receptor*—in the cell.) Testicular feminization is interesting because the extent to which it occurs can vary. In some cases such as described in (a), development goes in a 'feminine' direction: the child appears to be female, and so is reared as such, eventually acquiring a sense of herself as female. However, in other cases, some degree of masculinization of the genitals occurs before birth. Thus, the child is assigned as male, and brought up as male. What is interesting about these differing cases is that the child grows up to believe that he or she is the sex according to which he or she has been brought up.

If you thought that (b) and (c) did provide evidence for a hormonal effect before birth, you would not be alone, since this is precisely how many doctors and scientists have interpreted them. However, there may be other ways of interpreting the results. Look again at (b). Some scientists concluded that the display of excessive tomboyish behaviour indicated that the hormones prescribed to the mother had affected the fetal brain, and hence the child's behaviour.

☐   Suggest additional factors that are omitted by this explanation.

■   Two important ones are:

(a)   The alteration in the external genitals that resulted from the hormone exposure could have led the parents to behave differently towards the child than they might have done, had she not been so affected; in other words, they might have expected her to behave in 'masculine' ways, which in turn led her to do so.

(b)   A less obvious one is that children born to women with threatened miscarriages cannot readily be compared with children of normal pregnancies (as is implied in example (b)), since it is not immediately obvious whether any differences observed are due to the treatment, or to something about the abnormal pregnancy. Children born of pregnancies that did threaten to miscarry, but whose mothers were not treated with hormones, would provide a better comparison.

What of example (c)? This strange phenomenon occurs because of an enzyme deficiency in the body. Those affected lack the enzyme that converts testosterone to dihydrotestosterone. They are chromosomally male, but, because of the deficiency, their bodies lack the ability before birth to organize genital development in a 'male' direction. At puberty, however, the system is flooded with testosterone which pushes development into a 'male' direction. Scientists have often cited this effect as evidence of hormonal influence on the brain, arguing that, if social learning of gender role after birth were the only factor, then these people would find it immensely difficult to make the transition from one sex to the other. On the other hand, if their brains had been irrevocably affected by the pre-natal hormones, then they are simply making a transition for which their brains had, as it were, been biologically prepared.

This is a plausible explanation, but it is not the only explanation. The children having this enzyme deficiency are recognizable at birth. They are classified as female, as they have 'female type' external genitals. However, they are not quite the same as normal females and are locally referred to as guevedoces, meaning literally, 'balls at twelve'. Parents might thus bring up their child to expect to make the transition to 'being male'.

What then can be said of the significance of hormones in human development? They do seem to be important in the development of the reproductive tract and the external genitals. They may also influence the development of human behaviour in a similar way to that seen in other animals (Section 4.2.2). However, as the appearance of the external genitals at birth provides the criterion by which people are primarily classified as either male or female, and hence how they are raised, it is difficult to separate the effects of the hormones from the effects of the social environment.

Both chromosomes and genitals can be used to classify a person as one sex or another, and these are regarded as the biological sex of the individual. They usually match, but not always. A more rigorous division into one sex or the other would be based on whether the individual produced eggs or sperm (Book 1, Section 9.2.1), but such a test is rarely used on people. However, chromosomes and genitals do not determine gender; gender is socially defined. In Western society the wearing of trousers by either sex is acceptable; the wearing of dresses by males is not usually acceptable. The wearing of dresses has been ascribed as feminine, so people who wear dresses have a feminine gender. Not so long ago the same division applied to the wearing of jewelry, make-up and ear-rings, but currently these are worn to a greater or lesser extent by either sex. Clearly gender is a rather fluid concept. Psychologists recognize that people have a gender identity (that is, a person's self-concept), and a gender role (that is, the role in society that they either do play or are expected to play), but neither of these is dependent on biological sex.

These distinctions could be summarized as follows:

Biological sex	is principally classified into	female and male.
Gender identity	is principally classified into	woman and man.
Gender role	is principally classified into	feminine and masculine.

These are broad distinctions and they are sometimes used interchangeably, but, for most individuals, these three categories more or less map onto each other, and represent a distinction between two broad groups.

There is, in addition, a fourth category that is sometimes seen as related to the three that have just been outlined. This is *sexual preference*, which refers to the sex of the people whom one prefers sexually. Commonly, this is seen as a distinction between *homosexuals* (gay people), who have a sexual preference for people of the same sex and *heterosexuals* who have a sexual preference for people of the opposite sex. This distinction is often supposed to bear some relationship to the other distinctions, with the result that it is widely believed that someone whose sexual preference is not heterosexual must accordingly not 'fit' on one of the other categories. For example, the popular stereotype of lesbian is of a woman who (a) is likely to have a masculine role, (b) may have a masculine gender identity and (c) may not be entirely biologically feminine. Though some individual lesbians may think of themselves as less feminine/more masculine, there is in practice very little evidence to support this stereotype. Women who appear 'feminine' and 'masculine' are as likely among lesbians as among any other group of women.

In other words, although the other three categories more or less map onto each other, this fourth category of sexual preference does not bear any clear relationship to the other distinctions of gender. Nor is it easily fitted into two groups. Although some people are undoubtedly exclusively homosexual or heterosexual throughout their lives, there are also large numbers of people who fit in between, often at different points in their lives. And if pre-natal hormones do not determine gender identity or gender role, then they do not necessarily determine sexual preference.

Summary of Section 4.5

A short section such as this cannot do justice to the complexity of human development; that would require a course to itself. However, it has been possible to focus on the similarities between studies of human development and studies of development in other organisms. The main similarities are in the factors that influence development—genotype, external factors, hormones. Two genetic diseases were discussed. A number of pre-natal factors that alter development were presented, though in most cases their mode of action is not known. Visual acuity and hearing were touched on briefly to demonstrate that there are sensitive periods in these sensory modalities. Finally, the relationship between hormones and gender was discussed, which made a clear link with examples from earlier in the chapter (Section 4.4). The subject of gender raised the very important issue of social development in people.

4.6   Conclusion

The development of an organism is the result of the close interplay between the genotype of that organism, its internal environment and its external environment. This chapter has considered each of these three factors (genotype, internal environment and external environment) in turn to show how each of them can affect the development of behaviour. Each example in the chapter examined a different aspect of the developmental process, for instance, the process by which the genotype affects the structure of the brain, the process by which the hormone testosterone affects different brain structures, and the process by which visual experience affects the visual cortex. Each of the processes just mentioned was examined in a different species (the sphinx moth, the zebra finch and the cat, respectively). It was necessary to use different species, to illustrate different developmental processes, because those processes have been sought and examined in particular organisms and not others. For example, much of the human development section is descriptive rather than explanatory: thus, the effects of various teratogens were described, but no explanation could be offered as to how those teratogens exerted their effects. Such explanations must await studies on other organisms. The use of different organisms to study different developmental processes means that the study of behavioural development is patchy: a coherent story has yet to be told. However, if the assumption made at the beginning of this chapter is correct, that each example presented in this chapter illustrates a general process of development, then a picture of behavioural development begins to emerge, though clearly, for any particular organism, the picture is at present only an outline.

Objectives for Chapter 4

When you have completed this chapter, you should be able to:

4.1 Define and use, or recognize definitions and applications of, each of the terms printed in **bold** in the text. (*Questions 4.1, 4.3 and 4.8*)

4.2 Provide evidence that the genotype can influence the structure and functioning of the nervous system. (*Question 4.1*)

4.3 Discuss the significance of insect castes to an understanding of development. (*Questions 4.2 and 4.3*)

4.4 Give examples of non-specific factors and teratogens, and discuss their effects on development. (*Question 4.4*)

4.5 Explain what a sensitive period is, and show an understanding of how sensitive periods have been demonstrated with reference to the mammalian visual system and the auditory system of the owl. (*Question 4.5*)

4.6 Suggest reasons why the same factors might not exert the same effects on similar organisms during development. (*Question 4.6*)

4.7 Provide evidence that hormones influence the development and structure of the nervous system. (*Question 4.7*)

4.8 Explain how testosterone and oestrogen can sometimes have the same effect and sometimes a different effect on development. (*Question 4.7*)

4.9 Provide evidence that hormonal differences between juvenile males and females can result in both structural differences and behavioural differences in adulthood. (*Questions 4.8 and 4.9*)

4.10 Discuss the usefulness of the Tfm mutant in studies of sexual dimorphism. (*Question 4.9*)

4.11 Give a brief account of the development of sex and gender. (*Question 4.10*)

Questions for Chapter 4

Question 4.1 (*Objectives 4.1 and 4.2*)
What is a mosaic animal and why are they useful in the study of development?

Question 4.2 (*Objective 4.3*)
Suggest two ways in which animals of the same genotype can be physically very different as adults.

Question 4.3 (Objectives 4.1 and 4.3)
Explain how an initiator and a facilitator differ in their effects on the development of castes.

Question 4.4 (*Objective 4.4*)
From Figure 4.25, what is the vulnerable period for eye development as far as rubella infection is concerned? How does this vulnerable period accord with the general picture given in Figure 4.24?

Question 4.5 (*Objective 4.5*)
What is the significance of reverse-occlusion experiments to an understanding of sensitive periods?

Question 4.6 (*Objective 4.6*)
In Figure 4.25, some fetuses were unaffected by the rubella infection. Give three reasons why they might have been unaffected.

Question 4.7 (*Objectives 4.7 and 4.8*)
If very young female zebra finches are injected with either oestrogen or testosterone, and then as adults are injected with testosterone, they will sing. How can oestrogen and testosterone exert the same effect on the young female zebra finch?

Question 4.8 (*Objectives 4.1 and 4.9*)
Adult female canaries and zebra finches do not normally sing. If they receive an injection of testosterone, the female canaries do sing, but the female zebra finches do not. Explain this difference in terms of the activating and organizing effects of the hormone.

Question 4.9 (*Objectives 4.9 and 4.10*)
Which three of the following are correct?

(i)   The Tfm mutant produces testosterone and oestrogen.

(ii)   The Tfm mutant does not produce testosterone receptor.

(iii) The Tfm mutant does not produce testosterone.

(iv)   The SNB nucleus is found in male rats but not in female rats.

(v)   The HVc is of similar size in male and female zebra finches.

(vi)   The parietal cortex is known as the sexually dimorphic nucleus.

Question 4.10 (*Objective 4.11*)
Comment on the general statement that a boy brought up as a girl will have the gender identity of a woman.

References

Denenberg, V. H. and Whimbey, A. E. (1963).Behaviour of adult rats is modified by the experiences their mothers had as infants, *Science*, 142, pp. 1192–1193.

Gorski, R. A., Gordon, J. H., Shryne, J. E. and Southam, A. M. (1978) Evidence for a morphological sex difference within the medial pre-optic area of the rat brain, *Brain Research*, **148**, pp. 33–346.

Hotta,Y. and Benzer, S. (1976).Courtship in *Drosophila* mosaics: sex-specific foci for sequential action patterns, *Proceedings of the National Academy of Science, USA*, **73**, pp.4154–4158.

Knudsen, E. I. (1988) Sensitive and critical periods in the development of sound localization, in Stephen Easter, Kate Barald and Bruce Carlson (eds), *From Message to Mind*, Sinauer Associates, Inc. pp. 303–319.

Levine, R. B. (1986) Reorganization of the insect nervous system during metamorphosis, *Trends in Neurosciences*, **9**, pp. 315–319.

Munro, J. D., Sheppard, S., Smithells, R. W., Holzel, H. and Jones, G. (1987). Temporal relations between maternal rubella and congenital defects, *The Lancet*, **2**, pp. 201–204.

Schneiderman, A. M. and Hildebrand, J. G. (1985) Sexually dimorphic development of the insect olfactory pathway, *Trends in Neurosciences*, **8**, pp. 494–499.

Stein, Z., Susser, M., Saenger, G. and Marola, F. (1972) Nutrition and mental performance, *Science*, **178**, pp. 708–713.

Further reading

Arnold, A. P. and Gorski, R. A. (1984) Gonadal steroid induction of structural sex differences in the central nervous system, *Annual Review of Neuroscience*, **7**, pp. 413–442.

Bax, M., Hart, H. and Jenkins, S. M. (1990) *Child Development and Child Health*, Blackwell.

Mullen, P. H. (1983) *Handbook of Child Psychology, Vol II. Infancy and developmental psychobiology*, Wiley.

Nottebohm, F. (1989) From birdsong to neurogenesis, *Scientific American*, February, pp. 56–61.

Wiesel, T. N. (1982). Post-natal development of the visual cortex and the influence of environment, *Nature*, **229**, pp. 583–591.

CHAPTER 5
THE CELLULAR BASIS OF LEARNING AND MEMORY

5.1 Introduction

In Book 1, Chapter 6 of this course, learning was discussed in terms of behavioural modification due to experience. For an animal to store that learned experience, a memory for the event must have been formed (Book 1, Section 6.1). Memory can be defined as a change in the nervous system consequent on learning, by means of which information is stored; it is the inferred intervening process that connects learning to recall. Recall of memory may occur when there is subsequent exposure to the event that caused the animal to learn in the first instance; the change in an animal's behaviour which occurs as a result of learning may last days, weeks or longer—perhaps the entire life of the animal. Without recall of memory it is not possible to determine if the animal has learnt the experience, though as you read in Chapter 6 of Book 1, failure to demonstrate a change in behaviour on subsequent exposure to the original event does not mean that a memory has not been formed. There may just be a failure to recall; a forgotten telephone number was given to illustrate the question of whether the memory was unavailable, or was lost.

This chapter is concerned with the *cellular* basis of how memory is formed and stored in the nervous system of animals in response to different learning situations. The 'cellular' basis means the biochemical, physiological and anatomical changes within those cells and tissues of the nervous system which relate to memory storage processes for particular events. No complete answer can yet be given to the question of what constitutes the cellular basis of memory; research in this area is continuing at a rapid rate. So this chapter provides important generalizations based on an examination of several key models of learning.

Examples of memory formation will be given from a number of different learning models, from 'simple' learning in invertebrates through to more complex learning in mammals. It is not yet possible to resolve fully the issue of whether memory storage processes in these different animals are essentially similar events, or whether different forms of learning such as habituation, associative learning or complex learning (introduced in Book 1, Chapter 6) involve different mechanisms of memory formation and storage. Current opinion favours the idea that memory formation is not a single unitary process. However, one of the major problems in attempting to build a model (or models) of memory formation is that much of the work on cognitive psychology is carried out with humans or primates, whereas most neurobiological research on memory is done with other vertebrates and invertebrates.

One persuasive memory classification system, based largely on human and primate studies by Larry Squire in the USA, distinguishes between the existence of at least two separate types of memory systems or processes—declarative and non-declarative (which includes procedural memory)—both of which may be present in any one animal. (The concepts of declarative and procedural memory were introduced in Book 1, Section 8.2.2 and, along with other classification systems, were also considered in Book 2, Section 11.4.) Declarative memory, as the name suggests, is that which can be brought to mind as a proposition or a representation (that is, knowing 'that'), or, as the chapter on cognition put it, declarative memory is the ability to make links, to 'put two and two together'. Non-declarative memory includes procedural memory (that is, knowing 'how', such as the motor skill of riding a bicycle), simple classical conditioning and non-associative learning such as habituation and sensitization (Book 2, Figure 11.11).

☐　At what stage in the life of humans do you think non-declarative learning would have greater importance than declarative learning?

■　In the first two years of life, which is the time when the child learns how to coordinate movement and to deal with physical shapes. This is often called the *sensorimotor period of development*. It is interesting that declarative memories are almost never remembered from before the age of two.

The distinction between the two types of memory has an anatomical basis: in mammals, declarative and non-declarative memory appear to be organized in different parts of the brain. This distinction may also manifest itself when amnesia (loss of memory—Book 2, Section 11.4) occurs, or is induced experimentally in an animal or a person. In this case, loss of declarative memory does not necessarily mean loss of non-declarative memory. These differences point to the fact that memory is indeed far from a unitary process, even within a single species, let alone across different animal groups.

At the cellular level, however, most researchers believe that the many similar processes involved in memory formation are likely to be more notable than the differences. The reasoning for this is quite simply that because the basic mechanisms, such as those responsible for the action potential, are similar in all animals, it would seem likely that the cellular mechanisms involved with learning and memory will be the same (the parsimony principle of biochemistry, which is discussed more fully in Section 5.9). It is these cellular processes of memory formation which are the focus of this chapter. The difficulty comes when trying to relate the cellular mechanisms to specific behaviour patterns.

The chapter ends by considering the question of why there may be failure to recall memories.

5.2　What is memory?

One of the most striking features of memory is its very durability. People of 80 or 90 years of age can remember their childhood experiences, retained over a lifetime, during which every molecule of their body has been replaced many times. It was realized as long ago as the end of the last century that such stability must mean that when memories are formed, there must be structural changes in the brain—an alteration in the relationships between its neurons, which in some way could 'represent' the memory. But not all memorized items are retained for so long. Presented with a string of seven numbers at the rate of one a second, and immediately asked to repeat them back, most people can manage. If they are asked to repeat the sequence again half an hour later, most fail. However, if they are told that the numbers are important to remember, say a telephone number, then they are much more likely to succeed, and once they have crossed this half-hour barrier, are likely to be able to remember the number for several days or indefinitely if the number is used frequently.

This type of study led to the suggestion that there are at least two stages of memory formation: firstly an unstable, transient *short-term phase* (short-term memory, STM), which is a non-permanent phase. Secondly, there is a more stable *long-term phase* (long-term memory, LTM), a more permanent phase in which information is available for hours, weeks or years (Book 2, Section 11.4). Many cognitive psychologists now talk about working memory (Book 2, Section 11.4) instead of short-term memory, though here, because the terms are used largely in the context of a cellular explanation of memory formation, it is more appropriate to use short- and long-term memory. The concept of short- and long-term memory is probably relevant only to declarative memory. Non-declarative memory does not appear to be subject to interferences of the type that are described below, such as electro-shock, which blocks declarative memory.

In some forms of non-declarative memory, such as habituation and sensitization in *Aplysia*, which was examined in Book 1, Section 6.3.1, the cellular changes that occur during learning take place in a limited number of neural circuits, which are directly involved in the learning process. The neural processing that occurs in declarative memory, on the other hand, may involve several brain regions, not just those involved in the initial phases of memory formation. For instance, different brain regions may be involved in different aspects of the memory, such as the room in which an experiment on memory is being held, the colour of the test object in such an experiment, etc. Thus, in mammals, the hippocampus (Book 2, Section 8.8.7) plays a key role in short-term memory formation, but thereafter longer-term memory seems to involve other brain regions, most importantly the cerebral cortex.

There is no current consensus on the precise duration of short-term, as opposed to long-term, memory. To some extent it must be artificial to differentiate strictly between the two, because short-term memory is the process that must presumably precede long-term memory. Some workers have even defined an in-between stage called, appropriately enough, *intermediate-term memory*, though this will not be considered as a separate entity here. In some way the process of STM becomes consolidated into LTM, so, in reality, memory formation may be a continuum in which the short-term phase can merge into or even overlap with the long-term phase.

☐   Why might it be a good biological design if organisms evolved so that some items are lost from STM and gain access to LTM only if the organism is repeatedly stimulated or highly motivated?

■   It would ensure that only items of information having long-term significance are stored. This may be efficient since certain memory traces need only be held for a short time. Some items of information are outdated very quickly.

Note, however, that some memories are established immediately on first exposure to the stimulus, such as those acquired during imprinting (Book 1, Section 5.5).

☐ What neural events could occur so rapidly and transiently that they might code for STM?

■ Transient changes in the patterns of action potentials (Book 2, Section 3.8 and Book 3, Section 5.2.1) of particular neurons might be capable of coding for rapid short-term changes.

But it is difficult to see how these changes in the activities of neurons alone could be the basis of retention of memory for longer periods unless they set in train a series of events that alter the nature and perhaps number of connections between neurons (that is, alterations in neuronal circuitry). Such alterations would presumably alter their firing patterns in a permanent way. Thus, the idea has arisen, discussed at some length below, that longer-term memory must require more substantial modification of neurons and their interconnections. Before that discussion, the next section examines some of the reasons why there may be loss of information from STM.

5.3 Loss of information from short-term memory

5.3.1 Decay or interference?

Figure 5.1 shows a widely quoted result in psychology. Subjects are presented with three consonants (for example CHJ), forming what is known as a nonsense syllable and, following an interval, are asked to recall them. The length of the interval varies from trial to trial. If, during the interval, subjects are allowed to rehearse the letters (that is, repeat them to themselves aloud or silently), then, not surprisingly, recall is good. But, if rehearsal is prevented, performance deteriorates rapidly. To frustrate attempts at rehearsal, subjects are required to perform mental arithmetic (for example, they are asked to count backwards in threes from the number 741) in the interval between presentation and recall. It is argued that Figure 5.1 represents the natural decay of STM; decay is diminished or even prevented by rehearsal.

☐ Is this a valid conclusion to draw?

■ Perhaps. However, counting backwards in threes may actively erase the trace of the stimulus, the nonsense syllable in this case, and not merely thwart rehearsal. This may be how STM normally decays under conditions of continuous additional stimulation.

5.3.2 Disruption of STM

A different line of evidence that supports a distinction between the two types of memory comes from studies of the effects of traumatic interference (**trauma**) on brain function, such as accidental blows to the head. Such patients—after recovering consciousness—may not remember the events that led to the accident, or events for some considerable time prior to it, yet earlier memories are retained. Loss of memory for events immediately preceding trauma is called *retrograde amnesia* (Book 2, Section 11.4.2), and it points to a period of time in which memories are unstable, that is, when they have not become physically consolidated. In this indirect way, it permits an estimate of the time-course of the consolidation of trauma-sensitive memory into trauma-insensitive memory.

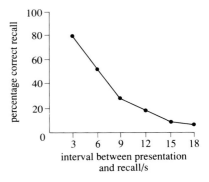

Figure 5.1 The decay of memory. A nonsense syllable such as CHJ is presented; the subject performs mental arithmetic for the intervals shown, and is then asked to recall the nonsense syllable. The percentage of trials in which recall is accurate is shown to fall very rapidly with time.

Patients suffering from trauma-induced amnesia do not provide very reliable evidence of the time-course of the consolidation. Early, experimental attempts to investigate short-term memory in rats in the 1940s and 1950s used electro-convulsive shock—a procedure analogous to that used therapeutically for some patients suffering from depression, called electroconvulsive therapy (Book 6). The results from the early experiments on rats, though, were far from clear. More recently, very mild, subconvulsive electric shock (equivalent to a tingle on the finger) has been found to disrupt short-term memory of certain learning tasks in chicks, a procedure that has yielded considerable insight into memory processes.

The day-old chick can be trained on a simple associative learning task called **passive avoidance learning**. This procedure was first introduced by Art Cherkin in California in 1969, and is used extensively by Steven Rose, Mike Stewart and their colleagues in the Open University as a means of studying learning and memory formation. In passive avoidance learning the chick is offered a small shiny bead to peck (Figure 5.2); the bead is coated with a distasteful, bitter-tasting substance such as methyl anthranilate, and, as a result of a single peck, the chick will avoid a similar, but uncoated, bead when offered one thereafter. It is said to be *passive* avoidance because the chick simply avoids pecking the bead after the unpleasant experience of tasting methyl anthranilate. The chick makes the association between the bead and the bitter-tasting methyl anthranilate. Control chicks can be offered beads coated with water, which they peck avidly, and then they subsequently peck the uncoated bead. The procedures used in these experiments are summarized in Table 5.1.

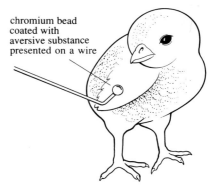

chromium bead coated with aversive substance presented on a wire

Figure 5.2   Passive avoidance training in day-old chicks using a shiny bead. See text for details.

Table 5.1   Summary of events in passive avoidance training.

Step	Experimental chick	Control chick
1	Chick is pre-trained to peck uncoated beads.	Chick is pre-trained to peck uncoated beads.
2	Chick then pecks bead coated with methyl anthranilate.	Chick then pecks bead coated with water.
3	Chick avoids pecking uncoated bead thereafter.	Chick continues to peck bead coated with water.

☐   What would be the importance of learning to avoid pecking at a bead coated with a noxious substance for a newly hatched domestic chick?

■   The importance of this behaviour pattern is that the chick has to learn quickly to discriminate between what is good food and what is not.

If a subconvulsive electric shock is administered within a minute of eliciting the disgust response for the bitter bead, the chicks become amnesic for the experience, and they peck the bead vigorously when it is presented again some time later. However, if subconvulsive electric shock is delayed for ten minutes after pecking the bitter-tasting bead, the chicks show recall and avoid the bead when subsequently presented with it. By altering the time of administration of the shock relative to the training experience, it is possible to alter the proportion of birds showing recall relative to those that are amnesic. If a population of trained birds is shocked five minutes after training, approximately half of them show recall, whereas the other half are amnesic when subsequently tested. (Chicks trained on a water-coated bead and subsequently shocked in a similar way do not show any changes in pecking behaviour.)

☐ What does this suggest about the time-course of consolidation of a short-term (that is, unstable) memory into a long-term (that is, stable) memory?

■ A possible interpretation of these data is that consolidation of information occurs within a period of ten minutes or so, although there is considerable variability between individuals.

These experimental data suggest a period of time for consolidation somewhat shorter than that found in people suffering from trauma. Further evidence for the time course of consolidation comes from experiments in which protein synthesis inhibitors are used. These drugs prevent the formation of proteins. If they are administered within minutes of a learning experience, the animals are amnesic for the learning experience.

However, the evidence from subconvulsive electric shock experiments can be interpreted in a different way, and may not simply show a disruption of STM and hence prevention of consolidation into LTM.

☐ Suggest another mechanism by which the chick data could be explained.

■ The subconvulsive electric shock could disrupt recall rather than consolidation. That is to say, the animal learns the task, a memory is formed, but for some reason the shock makes it unavailable as an expression in behaviour. The animal acts as though a memory has not been formed.

This interpretation receives some indirect support from reports of retrograde amnesia in human trauma victims, where it is a common finding that the length of the period affected by retrograde amnesia gradually decreases over time since the trauma. Memories that appear to be lost immediately after the trauma gradually come back over a period of weeks. This shows that the memories were present all the time but were unavailable; that is, retrieval was disrupted. (A similar problem with retrieval is sometimes encountered in everyday life when you know the answer to a particular question, but cannot recall it; the answer is 'on the tip of your tongue' and is usually remembered some time later.)

The principal conclusion from these studies is that memory processes—whether they be consolidation or recall processes—that occur immediately after a learning experience are fragile, and can be disrupted. Long-term memory processes, on the other hand, are very stable.

5.4 The nature and formation of long-term memory: consolidation of memory

The distinction between STM and LTM has proved useful as a means of examining the different facets of memory. Researchers are agreed that memories which are to be long-lasting must involve some kind of structural change; this is generally supposed to be a reorganization or change in effectiveness of connections between neurons. How such changes might occur, leading to permanent memory formation has intrigued investigators for a considerable time.

☐   Where might such changes occur?

■   The most likely place is at the synapse. To alter the effectiveness of synaptic transmission, changes could occur either presynaptically or postsynaptically. For example, there could be changes in the shape or size of the dendritic spines (see Book 2, Figure 2.8) and therefore in the associated synapses.

But how many synapses and neurons might code for a single memory, and is it possible to say that one memory association equals one synapse? Such a view would seem to be too simplistic. It is more likely that *groups* of neurons are in some way involved. It is difficult in any case to investigate single neurons and synapses amid the vast number in the brains of animals. What can be done is to take samples of tissue from brain regions likely to be involved in memory formation, and analyse the tissue for structural changes following learning. How such candidate regions of memory formation can be pinpointed will be discussed later in this chapter. First it is necessary to consider the background to current work, which has suggested that structural changes in the nervous system are indeed involved in memory formation.

The Spanish neuroanatomist Santiago Ramon y Cajal was the first to suggest, at the end of the nineteenth century, that the formation of new synaptic connections between neurons might be involved in the process of learning and memory storage. This was an idea involving great foresight because, before the electron microscope was invented, synapses were seen, even with the best light microscope that could magnify 1 500 times, as little more than tiny blobs of indeterminate structure. To discern any real detail of synaptic structure, one has to view very thin sections of brain tissue with an electron microscope at a magnification of at least times 30 000.

In 1949, Donald Hebb, a Canadian psychologist, proposed that activity within a given neural circuit leads to changes in the efficacy of the activated synapses in that circuit, so forming a 'neural representation' or 'memory trace' (sometimes referred to as an *engram*) of the acquired information. Re-activation of the trace retrieves the information concerning the event. This proposition was very astute, since most current models of learning have been concerned with activity-dependent changes in neural circuits, and, in particular, alterations in the efficacy of neurotransmission at synapses. There are a number of ways in which such modifications could occur, but it is instructive to consider Hebb's model of the cellular mechanism of learning, which remains the basis for much present-day thinking. His precise words are worth repeating here:

> When an axon of cell A is near enough to excite a cell B and repeatedly or persistently takes part in firing it, some growth process or metabolic change takes place in one or both cells such that A's efficiency, as one of the cells firing B, is increased.
>
> The most obvious and I believe much the most probable suggestion concerning the way in which one cell could become more capable of firing another is that synaptic knobs develop and increase the area of contact between the axon and dendrite or cell body. There is certainly no direct evidence that this is so…There are several considerations, however, that make the growth of synaptic knobs a plausible perception.
>
> D. O. Hebb (1949), *The Organization of Behaviour*, Wiley

Consider how it might work in practice (see Figure 5.3). The procedure is essentially that of classical conditioning. Consider an interneuron, C, on which two neural pathways (axons) converge. Suppose that the first, A, is a strong pathway, meaning that when A is active, C is always activated (Figure 5.3a), whereas the second, B, is a weak pathway, meaning that when B is active, C is rarely activated (Figure 5.3b). Suppose further, that, if A and B are active simultaneously, the activation of C by A results in some retrograde signal to B, such that the synaptic connection to B is strengthened; that is, C will have 'learned' to be active in response to B as well as A (Figure 5.3c). (Alternatively, the B terminals will have 'learned' how to activate C more effectively.) To give an example, suppose C is an interneuron that, along with others, plays a role in the process of salivation, A is part of a pathway from the olfactory system which activates neurons such as C when the smell of food is detected, and B is part of a weak pathway from the auditory system which responds to the sound of a bell. In a state of hunger, such a mechanism would then result in the animal now responding to the sound of the bell by salivating. The most obvious way that such strengthening of a connection could occur, said Hebb, would be if there were some growth or modification of the synapse connecting B and C.

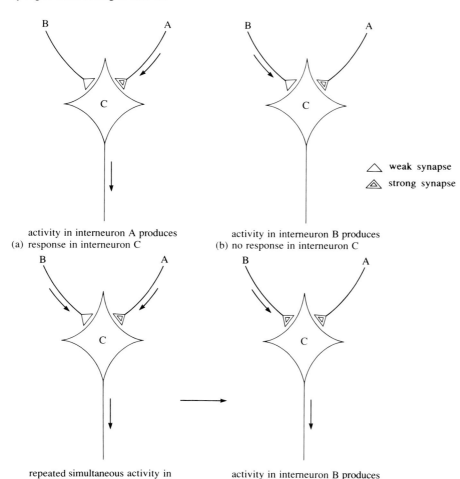

△ weak synapse
⬭ strong synapse

(a) activity in interneuron A produces response in interneuron C

(b) activity in interneuron B produces no response in interneuron C

(c) repeated simultaneous activity in interneurons A and B produces repeated response in interneuron C

activity in interneuron B produces response in interneuron C

Figure 5.3 Hebbian learning mechanism, showing converging neural pathways: (a) and (b) before learning, (c) after learning, when the pathway has been strengthened by persistent neural activity (see text for details). Three interneurons (A, B and C) are shown; in reality thousands will be involved in such a mechanism.

A number of variants on such **Hebb synapses** have been proposed in recent years, and they have been the basis of many attempts at neural modelling; in particular, theoreticians working with new types of computing, based on what is known as 'parallel distributed architectures' have envisaged the brain as containing many small assemblies of neurons, connected in a network of Hebb synapses, which can be strengthened or weakened in order to encode for new information, or to provide neural representations of the external world (that is, the interpretation of the world depends on learning what signals mean).

5.5   Structural changes in the nervous system following learning and memory formation

The synaptic changes proposed in Section 5.4 are not merely theoretical; they can be observed directly with both light and electron microscopes. Figure 5.4 shows a Golgi-stained neuron preparation. This staining technique (Book 2, Box 2.3) is capricious in that it only picks out a small proportion of the neurons in any section, but it stains each one in its entirety, showing the multiple branching dendrites, studded with dendritic spines. Each spine synapses with an axon terminal from presynaptic neurons not visible by this staining technique. It has been known for some years that rats reared in a so-called 'enriched' environment (that is, full of objects such as toys and obstacles, as described in Section 4.3.2) have substantial increases in branching of the dendrites compared with those reared in a 'restricted' environment (that is, a barren cage). There may also be increases in the density of dendritic spines (that is, numbers of spines per unit surface area of dendrite) on neurons in the cerebral cortex.

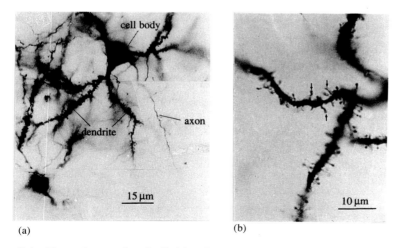

Figure 5.4   Photomicrographs of a Golgi-stained neuron from a P2 chick: (a) composite view, showing cell body, dendrites and axon; (b) higher magnification view of part of the dendrites, showing dendritic spines (arrows).

Recently it has been found that a single 'learning experience' can have similar effects. The Golgi preparation shown in Figure 5.4 is of a neuron from a region of the left hemisphere of the chick brain known to be involved in memory formation (the location of regions of the brain involved in learning in different animals is discussed in Section 5.10).

Twenty-four hours after passive avoidance training on a bead coated with methyl anthranilate (see Section 5.3.2), there are a number of dramatic changes in the cellular morphology (structure) of specific brain regions, notably the intermediate medial hyperstriatum ventrale (IMHV) and the lobus parolfactorius (LPO). (The locations of these areas are shown in Figure 5.5.) One of the most obvious changes is a 60% increase in the density of dendritic spines. This increase is seen when particular Golgi-stained neurons from trained birds are compared with similar neurons from a control animal that has pecked only a water-covered bead.

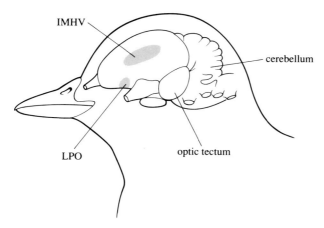

Figure 5. 5 Drawing of chick brain, showing the position of the IMHV (analogous to mammalian association cortex—Book 2, Figure 8.10) and the LPO (analogous to the mammalian basal ganglia).

Such an increase implies either the formation of new synapses or the relocation of existing ones from the shafts of the dendrites to the newly formed dendritic spines. Synapses on dendritic spines influence the postsynaptic neuron more strongly than those on the shafts of the dendrites, so relocation to the spines would have the required Hebb-type result of making the synapse more effective (Figure 5.6).

Changes in the dimensions of synapses can be seen in the IMHV and LPO regions of the chick brain. The electron micrograph of the synapse in Figure 5.7 shows some of the features that can be measured. Training the chick on the pecking task results in alteration in synaptic dimensions, the most notable of which is an increase in the length of the *synaptic apposition zone* (the active region of contact between the presynaptic and postsynaptic sides of the synapse, packed, on the postsynaptic side, with receptors for the neurotransmitters released from the pre-synaptic side; see Book 2, Chapter 4). How alterations in synaptic structure might occur is a matter for speculation. One very cogent suggestion is that synaptic activity following learning leads to alterations in adhesive interactions between the presynaptic and postsynaptic membranes, such that there is a change in synapse shape. In this process, glycoproteins (Section 2.4.3), which you will read more about in Section 5.8, are likely to be of central importance, as are changes in the concentration of extracellular calcium. The shape alterations result in an increased

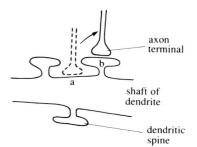

Figure 5.6 Relocation of a synapse from the shaft of a dendrite (a) to a dendritic spine (b). Synapse (b) is more effective in influencing the postsynaptic neuron than synapse (a).

surface area of contact between the presynaptic and postsynaptic membranes and a decrease in the synaptic cleft volume, events that could lead to enhanced synaptic efficacy.

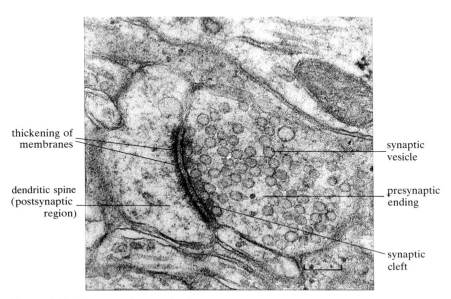

thickening of membranes

dendritic spine (postsynaptic region)

synaptic vesicle

presynaptic ending

synaptic cleft

Figure 5.7   Electron micrograph of a synapse in the IMHV of a chick. The presynaptic terminal, containing spheroid-shaped vesicles , makes synaptic contact on a dendritic spine. Note the thickened postsynaptic membrane. This is part of the synaptic apposition zone, which is thought to be the active zone of the synapse.(scale bar = 0.5 µm).

In the LPO, there is even an increase (up to 30%) in the numbers of synapses.

☐   Apart from increasing synaptic number in the area, by what other way could synaptic transmission be affected at synapses? (*Hint*   Consider the components of a synapse as shown in Figure 5.7).

■   One way would be by increasing the amount of neurotransmitter in the presynaptic terminal (for example by increasing the number of synaptic vesicles). This could lead to an increased release of transmitter across the synapse, and in turn to greater activity on the postsynaptic receptors, though the transmitter effect could occur in other ways.

Each synaptic vesicle contains roughly the same amount of neurotransmitter (Book 2, Section 4.2.3). It is therefore possible to measure the neurotransmitter content of synapses in a semi-quantitative way by counting the numbers of synaptic vesicles (this is called *semi-quantitative* because only a precise analysis of the composition of presynaptic axon terminals would give a true estimate of the amount of neurotransmitter present). Massive changes in the number of synaptic vesicles have been recorded: following training of chicks, a 60% increase in the numbers of synaptic vesicles in both the IMHV and LPO has been observed.

One might argue that morphological changes of the type described above could be caused by side effects of the learning process, such as stress, or could simply be due to the bitter effect of the unpleasant-tasting bead alone. The possibility exists that some or all of the morphological effects are not associated directly with the process of memory formation related to avoidance of the bead, but might be due to concomitants of the training procedure, for example, the perception of the taste of the methyl anthranilate.

☐   How could it be shown that the changes in synaptic and dendritic morphology are likely to be due specifically to the process of memory formation and not to any side-effects of the sort described above?

■   By comparing two groups of chicks, both trained on the passive avoidance task, but in one group the memory has been abolished by subconvulsive electric shock and in the other group the memory is present (Section 5.3.2).

☐   If the changes in synaptic and dendritic morphology are due specifically to the processes of memory formation, would the changes be present in one or both of the groups described in the previous answer? If one, which one?

■   The changes would be present in only one group—the group in which the memory is present.

Differences between the two groups still might not be due to memory formation alone. There is another possible explanation because the experiment contains a confounding variable.

☐   What is the confounding variable?

■   Subconvulsive electric shock. One group receives it, the other group does not.

To determine whether the increases in dendritic spine density were specific to memory formation processes, subconvulsive electric shock was administered to chicks five minutes after they had pecked the bead coated in methyl anthranilate.

☐   What is the significance of the five minutes?

■   It eliminates the confounding variable. Administering subconvulsive electric shock to chicks five minutes after they had pecked the bead makes some of them amnesic, but not others.

About 50% of chicks learn the passive avoidance response, receive the shock and retain the avoidance response (the *recall* group). The rest also learn the avoidance response and receive the shock, but are amnesic and peck the bead when tested (the *amnesic* group); see Section 5.3.2 if you are unclear about this.

The density of dendritic spines was estimated in neurons of the type shown in Figure 5.4. If the increase in density of dendritic spines previously observed was simply a consequence of the experience of tasting methyl anthranilate, the morphological changes should have been apparent in both the recall and amnesic groups. If, on the other hand, the dendritic spine increases were associated with memory storage for avoidance of the bead rather than the taste experience of methyl anthranilate, then the changes should have been present in the recall group, but not in the amnesic group.

Significant increases (with a mean value of 28%) in spine density were only observed in the recall group and not in the amnesic group (Figure 5.8). These results clearly demonstrated that the increase in spine density after passive avoidance training was specifically related to memory processes. In related experiments it was also shown that the chicks rendered amnesic did not show any difference in spine density compared with controls trained on a water-coated bead and then shocked, or indeed compared with untrained and unshocked chicks of the same age. The latter observation is important because it might have been argued that subconvulsive electric shock by itself causes structural changes in the tissue examined, but this does not appear to happen in chicks.

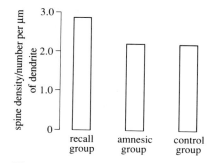

Figure 5.8   Spine density changes in dendrites of neurons from three groups of chicks, recall, amnesic and control; spine density is highest in chicks of the recall group.

Structural changes, at least in synapses (dendritic spine changes have not been investigated), have also been observed in the IMHV by Gabriel Horn and his colleagues in Cambridge. They studied a different type of learning in chicks, imprinting, which was considered in Book 1, Section 5.5. In this case, the main changes observed were alterations in the length of the synaptic apposition zone in the left hemisphere of chicks following imprinting on a flashing red light, compared with control chicks. In Chapter 4, data were presented to show that in swamp sparrows there is an increase in neuronal number in the higher vocal centre, which is associated with acquisition of song (Figure 4.21). This is complemented by synaptic alterations that include an increase in synaptic vesicle number, which could enhance synaptic transmission.

Such morphological changes, which have been found after other learning tasks in very different animal learning models—from habituation and sensitization in *Aplysia* to more complex learning situations in mammals—provide strong support for the idea that learning and memory formation involve some structural synaptic reorganization.

5.6   Biochemical approaches to the study of learning and memory formation

Memory formation can also be analysed in terms of changes at the biochemical level. Indeed, in the 1960s, there was a brief flurry of enthusiasm among biochemists and molecular biologists for the idea that somehow memories could be coded in the brain in terms of a specific unique protein, ribonucleic acid (RNA) or even DNA molecules. 'Hunting the molecules of memory' became a catch-phrase among memory researchers at that time. However, the improbability of this idea (and the implausibility of some of the experiments claimed to support it) soon led to its dismissal in respectable biochemical circles.

The two main biochemical approaches currently used in the investigation of memory formation are sometimes called *correlative* and *interventive*. In the former, an animal is trained on a particular task, and a search is then made for the biochemical changes that accompany the learning and memory formation. The problems with this approach are, firstly, that changes are likely to be small and localized, so it is desirable to have some idea in advance of the brain regions that may be involved, and, secondly, that all measurable learning tasks involve the animal in some type of motor or sensory activity, possibly stress and certainly arousal. It is therefore necessary to devise controls to ensure that observed changes are not merely the result of these necessary concomitants of memory.

The second, interventive, approach, endeavours to disrupt the process of memory formation by blocking a specific biochemical process, for instance with a transmitter antagonist (Book 2, Section 4.2.5) or an enzyme inhibitor, or by blocking a biosynthetic process. If, during learning, proteins are involved in synaptic construction or synaptic reinforcement, then a change in protein synthesis or breakdown may be detectable; the corollary of this is that blocking protein synthesis ought to interfere with memory formation. The best-known general example of this is the almost universal finding that administration of *protein synthesis inhibitors*, such as the antibiotics cycloheximide or puromycin, prior to a learning trial, or soon after, interferes with learning; it results in amnesia for the task, without affecting recall of already established memories. For example, injection of puromycin in mice 24 h after training to avoid a shock in a maze produces amnesia on re-testing of the animals at a later time. In controls injected with a neutral substance there is no impairment of memory.

☐ Would such experiments conclusively implicate protein synthesis in the process of consolidating memory?

■ Not necessarily. Firstly, considering protein synthesis, its blockage could hinder subsequent performance by a variety of general effects, such as altered enzyme levels or altered hormone production. Indeed, in large doses, such protein synthesis inhibitors are lethal. Secondly, the protein synthesis block, or other effects, could produce *retrieval* blocking or interference.

However, support for the role of protein synthesis in memory formation does come from studies in which it is possible to correlate memory formation with protein synthesis. It is possible to measure the rate of protein synthesis or breakdown by injecting radioactively labelled amino acids (the components of proteins) into the bloodstream (or sometimes by incorporating them into the animal's food). The animal will then construct proteins that are themselves radioactively labelled; differences in the rate of labelling between learning and control animals (not exposed to the learning task) indicate differences in the rate of protein use. The results of such studies suggest that structural changes do accompany learning, and involve elevated rates of protein synthesis, and also of glycoprotein synthesis. These experiments are discussed in some detail below.

5.7 Animal models of learning and memory

Training adult rodents to run mazes or press levers may require only relatively small-scale synaptic changes, likely to go undetected against the general level of cellular activity, which makes finding such changes akin to the needle in a haystack problem. It is therefore necessary to find more appropriate 'animal models' in which memory can be studied. Four models—two invertebrate, two vertebrate

—have proved particularly interesting in the last few years and features of these are discussed next.

5.7.1   Learning in *Aplysia californica*

The best-known invertebrate model uses the marine mollusc *Aplysia californica*, which has been extensively studied in the USA by Eric Kandel and his colleagues since the 1970s. *Aplysia* has a limited behavioural repertoire and a nervous system, which, at least in part, comprises large and readily identifiable neurons (sometimes misleadingly referred to as a 'simple' nervous system). In particular, *Aplysia* exhibits a number of bodily responses, notably the gill siphon withdrawal reflex (Book 1, Section 6.3.1). Withdrawal of its gill and siphon when touched by a rod or a jet of water can show habituation or sensitization, depending on the conditions. The set of neurons controlling this reflex can be removed and studied in isolation. Among them is a particular motor neuron with its associated sensory synapse. This single synapse preparation can itself be maintained in tissue culture, giving rise to what Kandel has called 'memory in a dish'. Although only a fraction of the entire system, this preparation has enabled him to analyse neural events related to behavioural phenomena.

For instance, in molluscs, one of the main neurotransmitters is serotonin; the amount released with each action potential in the presynaptic neuron alters during the process of habituation and sensitization. During habituation there is a decreased release of serotonin from the presynaptic terminal, whereas during sensitization an increased release occurs. This altered release is associated with dramatic membrane changes. In sensitization these probably include the opening of a particular ion channel (for calcium) in the membrane, and stimulation of the production of a second messenger (Book 2, Section 4.2.5). (The details are beyond the scope of the present chapter, but they are discussed in the books in the further reading list at the end of this chapter.) In addition, the biochemical changes following sensitization and habituation can be correlated with structural changes at synapses, notably an increase in the size of the presynaptic terminal and an increase in the numbers of synaptic vesicles in sensitized animals compared with habituated animals. These structural changes are reminiscent of some of those reported for the chick nervous system following passive avoidance learning (and indeed in mammalian learning models).

5.7.2   Mutant fruit flies

A second interesting approach has been the search for learning and memory mutants in that favoured organism of geneticists, the fruit fly *Drosophila melanogaster*. As genes code for proteins, some of which may play an important role in memory formation, it is likely that mutation in the genes that code for these proteins would disrupt the ability to learn and for memory for particular events. *Drosophila* can be taught a number of simple tasks, notably to avoid flying towards particular odours that are associated with electric shock. After subjecting a parent fly population to a mutagen (a substance that causes genetic mutation), flies can be discovered among their offspring, which, though not showing any other apparent behavioural deficits, cannot learn the odour/shock association. Several such mutant types, graced with names like *dunce*, *turnip* and *rutabaga*, have been identified. Interestingly, the biochemical effect of the mutations seems to be on the second messenger.

5.7.3 The vertebrate hippocampus

One of the most popular vertebrate systems to be studied in relation to learning and memory formation has been the rodent (especially rat) hippocampus. It is not known precisely what the hippocampus does in memory formation, but it is believed to play a key role in the processing of newly acquired information. The hippocampus is not itself the store, nor is it essential for retrieval of long-term memory. Rodents with hippocampal damage lack the capacity to create 'cognitive maps' (Book 1, Section 8.2.1), and cannot navigate in an environment in which they are dependent on external cues.

The hippocampus, with its layered neuronal structure (Book 2, Figure 8.30b) has thus long been of interest to neurobiologists because of the evidence of its involvement in both human and animal memory processing (Book 2, Sections 10.6 and 11.4). Humans with hippocampal damage are able to recover memories laid down before the damage occurred, but show a limited short-term memory capacity. However, they cannot transfer memories from short- to long-term store (believed to be somewhere in the cerebral cortex; see Section 5.10), and hence they deny knowledge of experiences occurring more than a few minutes previously. For example, if they have a conversation with someone who afterwards leaves the room and then returns a short time later, they will not recall that conversation. One of the best-known human examples was an unfortunate man in Montreal (code-named HM to prevent identification) who had a brain operation to relieve severe epilepsy. As a result of this operation, which removed most of both hippocampi (plus part of the amygdala), his epilepsy was cured. However, a side-effect was that he was only able to remember events that occurred prior to the operation, or current events, such as the conversation occurring at the moment. A short time later he would not remember the conversation.

☐ What process does this suggest HM lacked?

■ The process of consolidation.

HM was able to learn some things. For instance, he successfully learnt how to solve the manipulative puzzle called 'The Towers of Hanoi'. Each time he was presented with it he would say he had never seen it before, and yet each time his performance would improve (Book 2, Section 11.4.5)!

Because of the obvious ethical and practical problems of working with the hippocampus of humans or primates, most work on this structure in mammals has been concerned with rodents. In 1973, Tim Bliss and Terje Lømo in Oslo placed extracellular recording electrodes in rats, in the part of the hippocampus called the dentate gyrus (Book 2, Figure 11.8), and electrically stimulated the input pathway (the *perforant pathway*). A single electrical pulse administered to the input pathway produced a detectable response in the neurons of the dentate gyrus. However, after a train of high-frequency pulses to the input pathway, the response to a later single pulse was greatly increased. The increased response of the hippocampal neurons to a single pulse could be elicited several hours after a high-frequency

train of pulses. Thus, the high-frequency train of pulses changes the firing characteristics of the neurons in the dentate gyrus. This phenomenon is called **long-term
potentiation (LTP)**. Further experiments with rabbits have demonstrated LTP
lasting several weeks. LTP can even be induced in tissue culture, in slices of the
brain containing the hippocampus.

Long-term potentiation was instantly interesting to neurophysiologists, because it
is a physiologically generated stable change in neural properties; in this it might be
considered a model for (that is, the experimental equivalent of) memory. But
because of the special role of the hippocampus, might it not also prove to be a
mechanism for memory as well? There is evidence linking LTP to specific vertebrate learning tasks such as maze-running. It has been suggested that the hippocampus acts as a cognitive map, thereby playing a key role in spatial learning. One
of the most elegant demonstrations of this is the ability of rats to learn to find their
way around the Morris tank (Book 1, Section 8.2.1). The water in the tank is made
cloudy (by adding milk) and the rats are placed in it. They swim around at random
and after a time will accidentally find a submerged platform, on which they can
climb and so rest. A video camera mounted above the tank can track the route
taken by the swimming animal (Book 1, Figures 8.5–8.8). When placed back in the
water tank on subsequent occasions, and irrespective of position, the rat will swim
more or less directly to the platform. It is assumed that the rat forms a cognitive
map of its surroundings such that it remembers the position of the platform relative
to various cues in the room in which the tank is located.

☐   If the hippocampus is involved, how would you expect the rat's ability to learn
the position of the platform in the water tank to be affected by lesions to the
hippocampus?

■   Lesions to the hippocampus would dramatically impair learning of the location
of the platform.

And this is what happens. Further evidence of the role of the hippocampus in this
learning task is that the rat's ability to learn the position of the platform in the tank
can also be prevented by injection of drugs into the hippocampus. Most interesting
is the effect on learning of antagonist drugs (ones that act at the synapse to block
the action of the normal neurotransmitter). The antagonist AP5, for example, binds
to a specific sub-type of receptor for the neurotransmitter glutamate, namely the *N*-
methyl D-aspartate (NMDA) receptor. This receptor is believed to be activated in
the induction of LTP, and AP5 prevents the induction of LTP. (There is a
discussion of the mode of action of neurotransmitters in Book 2, Section 4.5.)
When the effects of AP5 on learning in the Morris tank were examined, the results
showed that AP5 does interfere with learning.

This dual effect of AP5 is illustrated in the following experiment. Each of 56 rats
was given an injection of AP5. The exact level of AP5 in the brains of the rats was
assessed after training in the Morris tank and after induction of LTP. The levels
ranged from zero to nearly $0.3\,\text{nmol}\,\text{mg}^{-1}$, but were essentially in four groups.
Over the next five days after the AP5 injection, each rat was given 30 trials in the
Morris tank, and the length of time the rat took to reach the submerged platform
was recorded on each trial. The mean time each rat took to reach the platform
(mean escape time) was calculated. Then an overall mean escape time for those

animals with similar levels of AP5 was calculated. These mean escape times are plotted against the level of AP5 in each group of rats in Figure 5.9a.

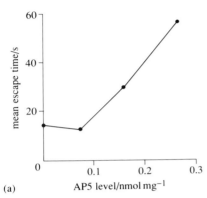
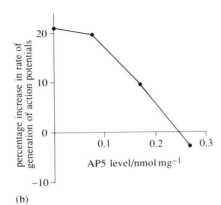

(a)
(b)

Figure 5.9 (a) Results from an experiment in which rats were injected with AP5 and then trained in a Morris tank. Mean time to reach the submerged platform (mean escape time) is plotted against the level of AP5 found in the brains of the rats. (b) The effect of AP5 on the induction of LTP. The percentage increase in the rate of generation of action potentials is plotted against the level of AP5. (The rate of firing of a group of neurons in response to a single electrical pulse is measured before and after LTP. The percentage increase in the rate of generation of action potentials after LTP is a measure of the effectiveness of LTP; the greater the increase, the more effective is the LTP.)

☐ In terms of mean escape times, how would the performance of rats that learnt about the tank compare with that of rats that learnt little about the tank?

■ Rats that learnt about the water tank would have shorter mean escape times than those rats that learnt little about the tank.

☐ What would you conclude from the evidence of Figure 5.9a about the relationship between AP5 and learning?

■ Essentially, as the level of AP5 in the brain increases, so learning ability decreases. Very low levels of AP5 (approximately $0.1 \, \text{nmol mg}^{-1}$ and below) appear to have little effect on learning ability.

These same rats were also used in a study of long-term potentiation, a study carried out after training in the tank, but before the levels of AP5 were assessed. This study showed that the higher the level of AP5 administered, the lower was the level of LTP induction. The results of this study are shown in Figure 5.9b.

These two sets of data show that levels of AP5 which interfere with learning also interfere with the induction of LTP. The inference from this is that learning in the Morris tank and the induction of LTP involve the same mechanism.

Studies of the biochemical mechanisms involved in LTP have provided evidence for two phases—an induction phase and a maintenance phase. The induction phase occurs in the postsynaptic neuron and involves a number of processes. The first of these processes is the depolarization of the postsynaptic neuron by the action of the neurotransmitter glutamate on non-NMDA glutamate receptors. This depolarization allows the second process to occur, namely the entry of calcium through the NMDA–glutamate-receptor-activated channels. The entry of calcium into the postsynaptic cell results in a cascade of biochemical events. It activates the second messenger system, and results in a series of cellular changes which could eventually lead to altered connectivity between the presynaptic and postsynaptic components, possibly including a messenger released from the postsynaptic cell which affects the presynaptic cell.

The maintenance phase is not fully understood, but is due in part to enhanced release of glutamate from presynaptic terminals. Much research on LTP is being carried out and, given the intensive effort of scientists working in this area, much new data will be gathered in coming years.

Whatever the nature of the biochemical events initiated by LTP, one would expect these changes to have structural correlates in the tissue of the hippocampus, and these have indeed been found. There have been reports of alterations in the shape of dendritic spines, an enlargement of the dendritic spine head area and an increase in the number of synaptic vesicles.

There is considerable similarity between the mechanism of LTP and that involved in the refinement of visual projections (Section 3.4.3). Both LTP and the refinement of visual projections reflect the ability of neurons to influence the current and future firing characteristics of other neurons. This influence is another manifestation of the plasticity of the nervous system.

5.7.4 Chicks

Passive avoidance learning, in which a young chick learns not to peck at a bitter-tasting bead, is a vertebrate model of learning, which was described in Section 5.3.2. Learning not to peck at the bitter bead initiates a series of alterations beginning with presynaptic and postsynaptic membrane changes, and proceeding by way of the production of proteins, to the lasting structural modification of these membranes. These synaptic modifications must in some way form the neural representations of the aversive bead-pecking experience and encode the instructions for the changed behaviour (avoid pecking a bead with these characteristics) that follows.

The brain regions that show enhanced neural activity as a result of training were first identified on the assumption that such increased neural activity would have a higher energy demand, that is, greater glucose utilization. The problem of how to measure this increase in glucose use was solved by 2-deoxyglucose (2-DG) auto-radiography (Book 3, Box 4.1) developed for the rat brain by Louis Sokoloff in the USA in 1977. This has now become a standard technique for determining changes in brain activity in a variety of animals (for example investigations of ocular dominance—Section 4.3.3).

By using 2-DG autoradiography in chicks that had been given passive avoidance training, it was possible to demonstrate increases in glucose utilization in the IMHV and LPO. These increases could be detected for at least thirty minutes after pecking at the bitter bead, and the later biochemical, physiological and structural changes were all localized to these regions. The IMHV, but not the LPO, is also implicated in filial imprinting (that is, imprinting onto a large moving object; see Book 1, Section 5.5) in the chick. These two regions appear to play different roles in memory. Lesions to either the IMHV or the LPO do not prevent chicks from learning the avoidance response. However, when chicks with such lesions were trained on the avoidance response and then tested a few hours later, those chicks with IMHV lesions were amnesic whereas those with LPO lesions could recall the avoidance response.

☐ What does this result suggest about the role of the LPO in memory?

■ The result suggests that the LPO plays no role at all: when there was a lesion to the LPO, the chicks learnt the avoidance response and could later recall it.

However, the following result shows that the LPO *does* play a role. The chicks were first trained on the avoidance response, and the LPO was then removed. When later tested, the chicks were amnesic for the avoidance response. So removal of the LPO *after* training renders the chick amnesic. A similar procedure in which the IMHV was removed after training was without effect. The results of these lesioning studies are difficult to reconcile with each other, but one possibility is that there is a 'flow' of memory from the IMHV to the LPO. In the absence of the LPO prior to training, the memory is retained in the IMHV.

The precise relationship between the site of memory formation and its storage is rather complex and not fully understood. There is even evidence that memory may not be stored solely in one region but may be distributed more generally throughout the brain. This is discussed in Section 5.10 in relation to work by Karl Lashley carried out in the 1930s.

Soon after training on the passive avoidance task there is an increase of both NMDA receptors and muscarinic receptors (a particular type of acetylcholine receptor). An NMDA receptor blocker such as the antagonist AP5 (which blocks the site without activating the receptor), produces amnesia, and chicks that have learned to avoid the bitter bead will peck it instead. This result has interesting parallels with that for LTP in the vertebrate hippocampus (Section 5.7.3). Within thirty minutes of training, there are other important biochemical changes, most notably in components of the second-messenger system. As would be expected, when chicks are injected with inhibitors of the second-messenger system, they develop amnesia for the passive avoidance task.

5.8 Long-term biochemical changes associated with memory formation

All the biochemical processes so far discussed have involved transient effects: receptor regulation, opening of membrane channels and the effects on second messengers. But, as discussed earlier in the chapter, long-term memory must involve more permanent changes to synaptic membrane structure, which requires the synthesis of new membrane proteins.

☐ How could new proteins be produced by the cells of the nervous system? (*Hint* What parts of the cell are vital for the production of proteins?)

■ To produce proteins, the cell must decode genes (Book 1, Chapter 3): the production of new proteins connected with memory formation requires that some additional genes are activated.

Thus, the biochemical mechanisms involved in triggering the synthesis of proteins after learning are assumed to involve the initial activation (that is, decoding) of particular genes. In fact the genes are believed to be of a type called *immediate early genes* (so called because they are the first to be activated). Immediate early-gene activation is believed to be initiated by signals emanating from the cell membrane. Thirty minutes after chicks have been trained on the passive avoidance task, immediate early-gene activation can be detected in both the IMHV and the LPO. Although the activation of these genes is sensitive to many types of sensory stimulation (including pecking at a water-coated bead), it has been possible, using carefully controlled experiments, to show that the activation is directly associated with learning the new task.

Whatever the intervening intracellular signals and genetic mechanisms, within 1 h of training there is enhanced synthesis of a variety of proteins destined for export from the neuronal cell body to its axon terminals and/or dendritic processes. Particularly relevant here are the glycoproteins of the synaptic membrane, because of the major role that several types of glycoproteins play in intracellular recognition and in stabilizing intercellular connections, especially during neural development (Section 2.4.3). For many hours following training of chicks, there is enhanced incorporation of radioactively labelled glycoprotein precursors into presynaptic and postsynaptic membrane glycoproteins.

It is presumed that the synthesis of the new membrane glycoproteins results in the presynaptic and postsynaptic structural changes that can be observed in Golgi-stained and electron micrograph pictures (Figures 5.4 and 5.7). If the Hebb model is correct, such changes must alter neural connectivity. In fact, electrical recording from the IMHV and the LPO in the hours after training reveals substantial increases in the firing rate in neurons in the IMHV and LPO. These increases are spontaneous and at a high frequency (*neural bursting*), a change in activity which is rather reminiscent of LTP.

5.9   Making memories—a summary

At the time of writing (1992), the studies described in Section 5.8 are still very much in progress. But they do at least provide a plausible mechanism by which the changes in neuronal connectivity which constitute the neural representation of associative memories can occur. What form might such a neural representation take? It is not clear if there is just one general biochemical sequence of events involved in all forms of learning, or many distinct sequences depending on the organism and the type of learning involved, such as association learning in *Aplysia*, maze learning in a rat or the multiple facets of human memory. If the *parsimony principle* by which biochemistry seems to operate holds as true in this case as it has done in others, then both broad universals and fascinating unique cases may be found. The parsimony principle can be most easily understood in terms of a particular substance, say dopamine. Suppose the biochemical events leading to the production of dopamine (that is, its synthetic pathway) are unravelled in one organism. If the parsimony principle applies, then other organisms that synthesise dopamine will use the same synthetic pathway. In other words, at the biochemical level, considerable commonality between organisms has been found.

As an example, consider a chick learning the passive avoidance task. The normal response to a bead must be to register the presence of an object of particular size and colour, and then to activate the motor responses of pecking at it. Experiencing the taste of the bitter bead must activate the biochemical machinery that will create in the brain some sort of model, against which any fresh presentation of an object can be matched. If the match is good, then the output response of pecking is inhibited. Although the initial storage of the model might involve sets of cells and synapses localized in one of the forebrain regions, the IMHV, it is probable that the bird's concept of 'bitter bead' is subsequently classified in several different ways, in multiple sites, including the LPO. The ultimate task of neuroscience is to discover how such cellular phenomena 'translate' into the experiences of conscious behaviour and memory; studies of the sort described above have a part to play in uncovering such translation rules, but it must be remembered that they provide only part of a much larger picture composed of the myriad complex interactions of the neurons and synapses of the brain.

A summary of the events involved in memory formation (Table 5.2), such as can be gleaned from the four models, considered in Section 5.7 is given below. This does not mean that every event occurs in each model; in some cases information on particular changes is not yet available.

Table 5.2 Events* occurring in short- and long-term stages of memory formation.

Stage of memory formation	Activity
short-term event (seconds to minutes)	altered neural activity (ion movements) as shown by increased glucose uptake in the brain, altered neurotransmitter receptor binding, of which the most notable is the glutamate receptor (sub-type NMDA)
short-term to long-term events (minutes to several hours)	activation of second-messenger systems, activation of immediate early genes, increased protein and glycoprotein synthesis; neural bursting (LTP-like effect)
long-term events (12 h or more)	structural changes including increases in synaptic number, increases in synaptic vesicle number, altered synaptic dimensions, changes in dendritic spine numbers and possibly in the pattern of neuronal branching

*To show that there is a continuum between the two stages, some events have been placed in a short- to long-term stage, though no discrete intermediate stage was discussed in the text.

5.10   The locus of memory formation

It was argued earlier that distinct populations of neurons are involved in memory formation. In mammals the hippocampus was said to play a key role in spatial learning, possibly through the mechanism of LTP. However, it is unlikely to be the case that the hippocampus is the memory store itself, because, as was described earlier with reference to humans with hippocampal damage, older memories are unaffected.

The IMHV and the LPO in the chick seem to be involved in memory formation, though memory may not be stored exclusively in these regions. Memory is impaired following lesions to these regions in the chick brain, but how much it is impaired depends on the timing of the lesion relative to the time of training. For example, in the chick IMHV a lesion given before passive avoidance training prevents consolidation (Section 5.7.4), but a lesion given 60 minutes after training has no effect on memory for the avoidance of the bead. This suggests that initial memory formation involves the IMHV, but afterwards memory 'relocates' in some way to other regions of the chick brain, one of which is the LPO.

The way in which sites of memory formation have been located in particular brain regions is usually by studies of the sort described in Section 5.7 with reference to the chick brain. Regions of altered neural activity in the chick brain following learning were located using 2-DG autoradiography. It is assumed that the change in glucose accumulation is related to increased neural activity associated with information processing in the early phases of memory formation (that is, STM). In living animals, changed brain function can be investigated using imaging techniques such as positron emission tomography (PET) (Book 2, Box 11.1) and magnetic resonance imaging (MRI) (see Box 5.1). Figure 5.10a is an example of an MRI scan.

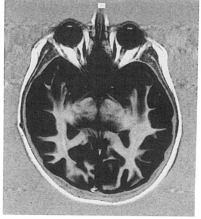

(a)

Box 5.1   Magnetic Resonance Imaging (MRI)

Magnetic resonance imaging (MRI) is based on the fact that each part of an object to which a strong magnetic field is applied absorbs radio waves of a particular frequency to varying degrees. The extent to which the radio waves are absorbed over a range of frequencies—the nuclear magnetic resonance (NMR) spectrum—can also be measured. By recording the NMR spectrum at different points within an object, such as a brain, an image can be created. Such images are usually obtained as two-dimensional 'slices' at particular depths within the object. In living tissue the amount of radiation absorbed, and hence the magnitude of the NMR signal, depends on the amount of water present, the extent to which the water molecules are free to move, and certain other, more technical, parameters. Bone, DNA and membrane lipids are not generally visible. The resultant image thus contains a great deal of information, and can be manipulated to highlight features of interest to the experimenter.

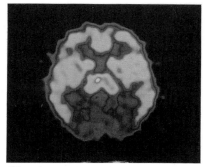

(b)

Figure 5.10   (a) MRI scan of human brain: the lighter the area, the less water it contains. (b) Positron emission tomography (PET) image of a human brain. The PET scan was made after dopamine receptors had been labelled with radioactive dopamine. The darkest regions on the photograph have the highest density of dopamine receptors.

A typical PET image, generated from studies on binding of a radioactive label to dopamine receptors, is shown in Figure 5.10b. The different intensities of dopamine labelling are often represented as different colours. PET images can be generated for a variety of receptors. Comparisons can be made of areas of low and high receptor densities between scans from normal control subjects and people or animals treated in different ways, or suffering from various diseases. Such comparisons reveal differences in neurotransmitter distribution and concentration.

By using PET or MRI brain scans, it is even possible to show how particular regions of the human brain are activated when a task is being learnt. However, the drawback of PET and MRI is that the resolution achieved is at best in the millimetre range, whereas the 2-DG autoradiography method used in rats and other animals achieves resolution in the tens of micrometre range (that is, 100 times greater), and therefore allows investigation of very discrete regions of the brain. But whichever of these methods is chosen to determine altered neural activity in a particular part of the brain, they do not alone prove that memories are stored solely in those regions, and the difficulty of determining precisely where memory is located in the chick brain is a good example of this.

So, are memories stored in one specific location in the brain or is memory more generally distributed? Suppose an animal is given a stimulus in a discrimination-learning task, for example having to discriminate between (that is, select) a cross that signals food and a triangle that does not signal food. Now, suppose brain tissue is surgically removed from the animal after it has learned the discrimination. Two possible sites of tissue removal are shown in Figure 5.11. In one case, neurons involved in the memory of the task are left intact and, in the other, part of the region involved in the task is removed.

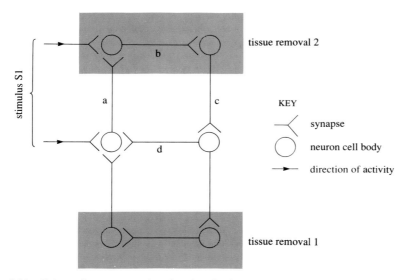

Figure 5.11 Schematic representation showing the four neurons (a, b, c, d) assumed to be involved in a particular memory, in this case the memory of the stimulus S1. See text for details.

☐ On the basis of Figure 5.11, what would be the expected effect of the two surgical interventions? (Assume the motor pathways remain intact.)

■ Tissue removal 1 might be expected to have relatively little effect, since the crucial neurons would still be intact. By contrast, tissue removal 2 might be expected to destroy the memory totally, since its loss would break the circuit.

In the 1930s, Karl Lashley found that rats with the cerebral cortex totally removed were unable to learn their way through a maze. It was therefore reasoned that the cortex is involved in such learning. Lashley trained rats to negotiate a maze and then removed bits of cortex to see how their ability was impaired. On the basis of Figure 5.11, you might expect some removals to have no effect and others to have a drastic effect. The site of the removal, as well as its magnitude, should be important. Lashley's findings are complex and sometimes interpreted in over-simplistic terms. However, it is sufficient to note that, for a task in which several sensory systems can be used, such as learning a route through a maze, the exact site of tissue removal was found to be unimportant. It is not the case that removal of one small specific region of cortex has a large effect, whereas removal of another area has relatively little effect. In general, the more cortex that is removed, the poorer the performance. This usually leads to the conclusion that memories are established diffusely throughout the cortex. For example, circuits 1, 2 and 3 in Figure 5.12 carry the memory of the stimulus. Normally, all three may act together, but if, say, 1 is knocked out, it still leaves 2 and 3, which may carry sufficient information to enable the rat to negotiate the maze. In other words, memory shows redundancy: it incorporates a 'back-up' system.

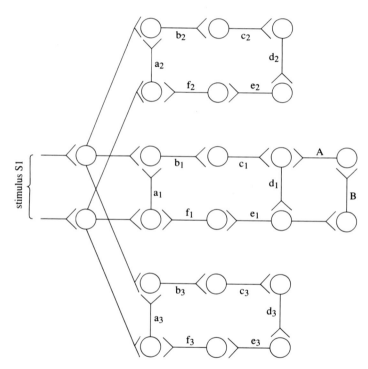

Figure 5.12   Memory of stimulus S1 is shown as coded in the form of a_1, b_1, c_1,... and also as a_2, b_2, c_2..., etc. Each of these circuits carries the same information. Hence any given memory is replicated several times throughout the cortex. This inbuilt redundancy should make the memory less vulnerable in the event of brain damage. (By the same logic, people working on word processors make several computer back-up copies of their work in case one is lost.) Note the different memory A, B, d_1, e_1, which shares the neurons d_1, e_1. A given neuron may participate in many different circuits and hence different memories.

It is possible to envisage two kinds of redundancy. The rat may use more than one sensory modality to negotiate a maze (for example touch and vision). This would provide two compatible memories for the maze. In addition, a particular memory (for example visual memory) may be widely distributed throughout the brain. If the maze is such that rats use more than one sensory modality to negotiate it, the site of tissue destruction is less important than if only one sensory modality is used. If, say, only vision is employed and parts of the cortex having a primarily visual function are removed, the effect is particularly destructive.

Interestingly, there are numerous reports suggesting precisely the opposite interpretation of memory from the idea of diffuse representation derived from Lashley's results. Some brain surgery is done on conscious patients, using only local anaesthesia on a part of the head; the brain itself has no receptors that can detect tissue damage—that is, nociceptors—and requires no anaesthesia. When specific points on the cortex of some of these conscious patients were stimulated, they reported that the stimulation evoked specific memories. For example, one man saw himself in his childhood home laughing and talking with his cousins. Repeated stimulation of the same part of the cortex evoked the same memory. However, it is not known whether the same memories could have been evoked from other regions of the brain, since the cortical stimulation was incidental to the surgery, and therefore was not directed in a systematic way.

Summary of Chapter 5

Memory is the change in the nervous system consequent on learning, by which information is stored; it is the inferred intervening process that connects learning to recall. Memory formation is probably not a single unitary process for all learning events, even within a single species, let alone across different animal groups. There are two forms of memory: declarative memory includes the ability to know *that*—that is, to put two and two together—whereas non-declarative memory, which includes procedural memory, is concerned with knowing *how* to do things, such as motor skills.

There are at least two stages of memory formation, a transient short-term phase (short term-memory, STM), lasting for up to several hours, followed by a more stable phase (long-term memory, LTM). The concept of short- and long-term memory is probably relevant only to declarative memory. Trauma and subconvulsive electric shock result in loss of memory for events immediately preceding trauma or shock—retrograde amnesia—pointing to a period of time in which memories (STM) are unstable. Formation of long-term memory (consolidation) involves structural alteration that results in change in the effectiveness of connections between neurons. This is most likely to occur at the contact zone between neurons, the synapse. This idea owes much to the work of Donald Hebb. Structural changes in the nervous system following learning and memory formation can be observed directly at both light and electron microscope level. Biochemical approaches to studying learning and memory formation are concerned with the identification of the sequence of biochemical processes that must be involved in the short- and long-term modification of synaptic structures and, therefore, connectivity between neurons. In correlative approaches to such studies, animals are trained on a particular task, and a search is made for the biochemical changes that accompany the learning and memory formation, whereas interventive approaches disrupt the process of memory formation by blocking a specific biochemical process, for example by administration of protein synthesis inhibitors, such as the antibiotic puromycin. This interferes with learning, resulting in amnesia for the task, without affecting recall of already-established memories, indicating that the blocked biochemical process was necessary for the memory to be formed.

Four animal models for the study of memory were considered: (a) *Aplysia californica,* (b) mutant fruit flies, (c) the vertebrate hippocampus, and (d) chicks. The cellular changes that occur in the hippocampus and passive avoidance learning in chicks were considered in some detail. The hippocampus plays a key role in formation of long-term memory, but is not itself the store of memory, nor is it essential for retrieval of long-term memory. Humans with hippocampal damage are able to recover memories laid down before the damage occurred, and show a limited short-term memory capacity. However, they cannot transfer declarative memories from short- to long-term store (believed to be somewhere in the cerebral cortex).

The phenomenon of long-term potentiation (LTP) in the hippocampus was examined. LTP may prove to be not only a model but also a mechanism for memory, because there is evidence linking LTP to specific vertebrate learning situations such as maze-learning by rats. A specific sub-type of receptor for the neurotransmitter glutamate, the NMDA receptor, plays a crucial role in LTP. Following induction of LTP, there are also morphological alterations in hippocampal tissue.

Among the long-term biochemical changes associated with memory formation are enhanced synthesis of a variety of proteins destined for export from the neuronal cell body. Particularly relevant are the glycoproteins of the synaptic membrane, because of the major role that several types of glycoprotein play in intracellular recognition and in stabilizing intercellular connections, especially during neural development. Such long-term changes are assumed to begin with the activation of immediate early genes.

Memories are not necessarily stored in one specific location. Following brain lesion studies in mammals, Lashley found that the exact site of removal of brain cortex was not very important. In general, the more cortex that is removed, the poorer the performance. Memories are established diffusely throughout the cortex. If one part of the brain is removed in a rat, the remainder may still carry sufficient information to enable the rat to negotiate the maze. In other words, memory shows redundancy: it involves a 'back-up' system.

Objectives for Chapter 5

When you have completed this chapter, you should be able to:

5.1 Define and use, or recognize definitions and applications of, each of the terms printed in **bold** in the text.

5.2 Define, and distinguish between learning, memory and recall. (*Question 5.1*)

5.3 Explain why it is thought that memory formation may not be a unitary process. (*Questions 5.2, 5.3 and 5.4*)

5.4 Explain the meaning of short- and long-term memory. (*Question 5.5*)

5.5 Discuss the term *consolidation* in the context of memory formation, and describe how the use of subconvulsive electric shock has contributed to our understanding of the different phases of memory formation. (*Questions 5.3 and 5.4*)

5.6 Explain, using invertebrate, bird and mammalian examples of learning, why permanent biochemical and morphological changes in the nervous system might be expected to accompany learning, and discuss the temporal relationship between the changes. (*Questions 5.6 and 5.7*)

5.7 Describe the possible cellular basis for short- and long-term memory. (*Questions 5.6, 5.7 and 5.8*)

5.8 Describe Lashley's work, and the concept of neural redundancy and its implications for memory models. (*Question 5.9*)

Questions for Chapter 5

Question 5.1 (*Objective 5.2*)

In the experiment shown in Figure 5.1, suppose it were found that subjects do worse at the task if, a few minutes before testing, they are given another similar memory test. Would this argue for or against the idea that memory decay underlies the recall failure?

Question 5.2 (*Objective 5.3*)

In mammals with damage to the hippocampus, certain simple motor skills can be learnt, but more complex learning such as that normally shown in the Morris tank cannot occur. Why should this be so?

Question 5.3 (*Objectives 5.3 and 5.5*)

A brain-damaged woman in North America showed the following characteristics. She could remember her name, early childhood, the assassination of President Kennedy, etc. She could conduct a normal conversation. However, although her doctor met her every day, she showed few signs of recognizing the doctor or remembering anything from the previous day's consultation. In terms of consolidation and retrieval of memory, in what way is she abnormal?

Question 5.4 (*Objectives 5.3 and 5.5*)

Consider the case of the woman described in Question 5.3 and examine Figure 5.1. Can one really say she might have normal short-term memory but that consolidation might be a problem?

Question 5.5 (*Objective 5.4*)

What evidence is there that non-human animals have short-term memory?

Question 5.6 (*Objectives 5.6 and 5.7*)

After injection of substances that inhibit protein synthesis, which of the following might be expected?

(a) Neuron activity would be hindered because of increased potassium ion permeability.

(b) Memory formation would be prevented because protein synthesis inhibitors cause excessively increased neurotransmission across the synapse.

(c) Neuronal firing would be increased, making consolidation easier.

(d) The formation of the morphological changes in the nervous system that constitute long-term memory would be hindered or prevented.

(e) Short-term memory would be obliterated.

(f) The transfer of each short-term memory to its corresponding specific stable molecule code would be prevented.

Question 5.7 (*Objectives 5.6 and 5.7*)

On the day of hatching (day 1) domestic chicks were trained to respond to a particular tone (an acoustic imprinting task). One week later, an examination was made of tissue from a region of the chick brain known to be a key centre for learning and memory. A decrease in the density of dendritic spines on neurons in this area was found, compared with the density of spines in the same region of newly hatched chicks. This was claimed as conclusive proof that long-term memory formation following acoustic imprinting in chicks is caused by a process of synaptic elimination. (a) Comment on the plausibility of this assertion, and consider whether a change in dendritic spine density is a precise measure of synaptic number. (b) Suggest adequate control experiments that should have been performed to demonstrate that the changes in the density of dendritic spines is directly related to acoustic imprinting.

Question 5.8 (*Objective 5.7*)

The synaptic transmitter acetylcholine is normally inactivated by AChE (acetylcholinesterase). The substance DFP blocks the action of AChE, and hence inhibits the breakdown of acetylcholine. Suppose injection of small doses of the blocking substance at the time of learning was shown to speed up learning. Suggest a mechanism by which this effect could occur.

Question 5.9 (*Objective 5.8*)

A rat is trained to negotiate a maze. A small amount of brain tissue is then removed and it is found that the rat is unaffected in its ability to negotiate the maze. Which of the following conclusions can you draw?

(a) The rat is using several different sensory modalities in finding its way through the maze.

(b) The rat is using only one modality but the memory is spread diffusely.

(c) The rat may be using more than one modality, and any memory may be diffusely located.

(d) The removed region had absolutely nothing to do with the task that the rat had learnt.

(e) Any one of three things must be true, namely that the region has nothing to do with the memory, or that the rat is using several different modalities and/or the memory associated with a given modality is diffusely located.

Further reading

Dudai, Y. (1989) *The Neurobiology of Memory: Concepts, Findings, Trends*, Oxford University Press, Oxford.

Squire, L. R. and Lindenlaub, E. (1990) *The Biology of Memory*, Symposia Medica Hoechst 23, Schattauer Verlag, Stuttgart and New York.

Shaw, G. L., McGaugh, J. L. and Rose, S. P. R. (1990) *Neurobiology of Learning and Memory*, Advanced Series in Neuroscience, Volume 2, World Scientific, Singapore.

EPILOGUE

The purpose of this epilogue is to reflect on the book as a whole, and hence on development and learning.

Development has been considered from a restricted perspective: the development of the entire organism, its multiplicity of different cell types, its patterns and organization of structures has not been addressed. Rather, the focus of this book has been exclusively on the nervous system, and this epilogue is similarly focused.

During development there is a high degree of change, involving increases in neuronal cell numbers (followed by decreases), increases in the numbers of cell processes (for example, dendrites and axons), followed by decreases and increases in the lengths of these processes. As the organism matures, so the rates of change in these things decrease, until, in the mature organism, they are minimal. There are, though, considerable changes at the level of the synapse in both the young and the mature organism. A distinction can therefore be made between the gross anatomy of the nervous system, which, once established, is relatively stable, and the micro-anatomy of the nervous system, for example, synapses, where changes can and do occur throughout life. This distinction is important when considering the principal theme of this book, plasticity.

The growing nervous system can exhibit the properties of plasticity, but generally the gross anatomy of the nervous system is similar between individuals of the same species. Thus, altering the position of guide-post cells alters the route taken by growing axons, or altering the input to the antennal lobe affects whether a macroglomerular complex is formed. But these are manipulations that most individuals do not encounter. Normally, neurons make connections with targets in a predictable way. Later in life, though, there are no spontaneous changes in the gross anatomy (even the changes that occur in response to injury are fairly localized). In most species the nervous system loses completely the ability to form new pathways or new structures during the period referred to as maturity. There are notable exceptions. Thus, the higher vocal centre of the canary increases in size each spring. These exceptions aside, the general rule appears to be that, once established, the gross anatomy of the nervous system is stable. Put another way, the gross anatomy of the nervous system is plastic during development but stable in maturity. Development then is not a process that ceases at maturity; rather, the capacity for change in the nervous system simply decreases with age.

Another aspect of the young, developing nervous system is that the changes seen in gross anatomy are adaptive; there is a 'tuning' of neuronal connections. This tuning is in contrast to the changes in gross anatomy seen in the ageing nervous system, where the changes are not adaptive, for example, there is a decrease in length of dendrites in the ageing nervous system. The period of development, then, can be distinguished from the periods of maturity and old age by rapid adaptive change in gross anatomy.

The situation is entirely different for the micro-anatomy. Changes at the level of the synapse occur throughout life; in the young animal there is competition between axons for survival factor, with the loss of inappropriate synapses. In the mature animal there are changes in the neuromuscular junction and changes at the synapse commensurate with learning; in the old animal, changes associated with learning continue. Thus, micro-anatomy remains plastic throughout life.

Three further features need to be drawn out of the continuing plasticity of the nervous system.

First, there is considerable similarity between the processes involved in the fine-tuning of connections between neurons during development, and the processes involved in learning. The model developed for the fine-tuning of synapses during development is that functional synapses are strengthened, whereas non-functional synapses are lost. Function in this context requires that both the presynaptic terminal and the postsynaptic terminal are active simultaneously. This model is remarkably similar to that developed by Hebb to account for learning, which requires the concurrent activation of both presynaptic and postsynaptic cells. The similarity of the models is supported at the biochemical level, because in both processes the NMDA receptor has been implicated. Indeed it has been suggested that the NMDA receptor is a coincidence detector, being active only when both presynaptic and postsynaptic cells are active.

The second feature is that the changes proposed to account for learning cannot occur until synapses have been formed. Synaptogenesis occurs at different rates and at different times in different parts of the brain (in humans, much of this synaptogenesis is post-natal). It follows that the ability to learn has a time course that lags behind synaptogenesis. For an infant to remember its mother's smell or face, synaptogenesis must already have occurred in the appropriate part of the brain. This leads on to the third feature.

Memories formed in infancy (the argument applies to all memories, but is more poignant for childhood memories) can be retained throughout life, despite continued synaptogenesis and changes in micro-anatomy occurring during this period. There is then a conflict between the idea of the plasticity of synapses developed above and the durability of strengthened synapses formed as a result of learning that gives rise to long-term memory. Perhaps, though, only those synapses that have not been strengthened retain plasticity. In addition to the changes in the mature nervous system that are adaptive, there may also be changes that are maladaptive, resulting in disease.

Finally, it is one thing to show changes in the nervous system or changes in behaviour following an environmental change, but it is quite another to be able to make predictions. The difficulty of foreseeing the consequences of environmental factors on the nervous system and on behaviour arises in part from the balance between resilience and plasticity, and in part from the complex relationship between the nervous system and behaviour.

GENERAL FURTHER READING

Ribchester, R. R. (1986) *Molecule, Nerve and Embryo*, Blackie.

Wolpert, L. (1991) *The Triumph of the Embryo*, Oxford University Press.

ANSWERS TO QUESTIONS

Chapter 1

Question 1.1
Cell divison in the brain occurs in the walls surrounding the ventricles, where stem cells divide to produce neurons and glia.

Question 1.2
(a) The line stops at about age 20 months.

(b) No the brain has not stopped growing; it is still increasing in size, though the *rate* of increase by 20 months is low.

Question 1.3
Neuronal plasticity means that the connection between a neuron and a target is neither specified nor fixed, whereas neuronal specificity means that a particular neuron is always connected to a particular target.

Chapter 2

Question 2.1
Because only at a particular stage of development are both tissues at an appropriate state of differentiation, the induction of a secondary neural plate will only occur if the grafted tissue is capable of signalling to the belly ectoderm and the belly ectoderm is capable of responding to the signal. Throughout development the results of cellular interaction depend on the state of differentiation of the interacting cells. Both of these abilities are dependent on the state of differentiation of the graft tissue and the host tissue; these states of differentiation continuously change with time.

Question 2.2
In animals having a very small number of cells (such as the nematode worm), cell lineage or ancestry can be investigated by simply following what happens to identified cells during development. In more complex animals, injections of long-lasting dyes into the embryo at an early stage give this kind of information.

Question 2.3
(a) In the cerebral cortex, thymidine autoradiography shows that cells move from where they were born to occupy particular layers.

(b) In the peripheral nervous system the migration of neural crest cells (some of which form the autonomic ganglia—Section 2.3.5) has been investigated by grafting cells that can be identified into animals at various stages of development, and watching where they end up.

Question 2.4

Induction involves the interaction between cells or tissues. One tissue must produce a messenger of some sort, which the other tissue can not only detect through a receptor, but can also respond to.

Question 2.5

In normal mammals the later-born neurons migrate past the earlier-born neurons to occupy the upper cell layers of the cerebral cortex. In the reeler mutant mouse the migration is abnormal, so that early-born neurons occupy superficial instead of deep positions. Despite their abnormal location, they still differentiate into pyramidal cells (characteristic of the deeper cortical layers), whose axons project to other brain areas, suggesting that for these neurons, differentiation is more affected by birth date than by cell position.

Question 2.6

There is abundant evidence that axons do not grow straight to their targets. They constantly make and retract branches, 'searching' for the appropriate substrate or chemical cues along which to grow (Figure 2.14). Thus, the movement of a growth cone is hesitant. They also deviate around other cells (Figure 2.18), and they do not grow through tissue that is too dense.

Question 2.7

Chemotactic factors affect axonal growth through direct contact. Direct contact between the growth cone and signalling tissue is not necessary for chemotropic factors such as 'maxfactor'.

Question 2.8

The interaction between growing axons and their environment is extremely complex and varied. The cell membrane has different receptors at different times and in different regions. It is highly unlikely that axon guidance could depend on one factor, just as it is rare to use only one visual cue to reach a friend's house. Such a system would also be very inflexible.

Chapter 3

Question 3.1

Selective cell survival is responsible for matching the numbers of neurons to the size of their targets, wherever these are situated. The effect of a chemotrophic factor can be very precise, since it will only affect innervating neurons and does not affect nearby neurons innervating other targets. Thus, only a small number of chemotrophic factors can control the sizes of a number of innervating neurons.

Question 3.2

The most obvious reason would be that not all the motor neurons established make synaptic contact with the muscle and so would not survive. Another possibility might be that, even if initially all neurons made contact with the muscle, there is still not enough target muscle to supply all the motor neurons with enough chemotrophic factor.

Question 3.3

1   The factor must be present in the target tissue during the period of cell death.

2   The factor should be taken up by axon terminals and transported back to the cell body of the innervating neurons.

3   The factor should be able to increase the survival of neurons in the living animal and in tissue culture.

In many cases it is not possible to establish all these criteria. In addition it may have also have occurred to you that the blocking of the action of the factor with antibodies should result in an increase in neuronal death in the living animal and in tissue culture.

Question 3.4

This suggests, in common with many other developing systems, that the mechanisms responsible for the normal development of this reflex pathway are only present just before metamorphosis. Thus, if the operations were performed just before metamorphosis, the mechanisms would be disrupted and formation of the reflex pathways would be prevented.

Question 3.5

The mechanism of functional tuning allows the development of highly precise connections in the visual system, which would be difficult to achieve by other means. Specific guidance of axons does produce an organized projection, but without the precise details seen after functional tuning has occurred.

Question 3.6

The fish eye continues to grow throughout life. In mature animals the retinal axons are therefore provided with a suitable environment for axon growth and guidance cues. Also no retinal ganglion cells die when their axons are damaged, so that all retinal axons have an opportunity to re-grow to their targets.

Question 3.7

The glial cells in peripheral nerves support axon growth, whereas those in the central nervous system do not. Since peripheral nerve grafts do not provide guidance cues, the chances of axons re-innervating their original targets are still very slim.

Question 3.8

Immature glial cells appear to promote axon growth. Also, the fact that axons will grow through peripheral nerve grafts suggests that glial cells surrounding the peripheral nerves (Schwann cells—Book 2, Section 3.8) facilitate axon growth too (Section 3.5.4).

Question 3.9

Most anatomical methods involve looking at dead, fixed tissue and therefore do not reveal subtle changes in structure that might be occurring in a living animal. Until electrophysiological methods were developed there was no method sensitive enough to detect changes in activity of cells that might represent changes in connectivity unless such changes were very obvious.

Question 3.10

No. The experiments of Pat Wall (Section 3.6.2) suggest that, at least in the spinal cord, the accuracy of the projection is a result of the inhibition of activity ('silencing') of inappropriate connections rather than their loss.

Chapter 4

Question 4.1

A mosaic animal has some cells of one genotype and other cells of a second genotype. Mosaics are important in the study of development because they provide evidence for the involvement of a particular genotype in the development of particular structures.

Question 4.2

There are numerous possible answers to this question. Using examples from the castes of ants, there could be non-isometric growth, or the animals could be subject to different facilitators, initiators or predisposers. You could also point out that exposure to different teratogens could result in different body forms, as could exposure to different hormones.

Question 4.3

A facilitator enhances a property that is present in the organism anyway. Thus, food is a facilitator of growth. An initiator has to be present for a particular course of development to occur. Thus, ants developing at high humidity become soldiers, say, whereas those developing at lower humidities become workers.

Question 4.4

Rubella infection between the ages of 3 weeks and 12 weeks inclusive affects eye development. Therefore, the vulnerable period is weeks 3–12 inclusive. From Figure 4.24 the vulnerable period for eye development is about weeks 4–9 inclusive. Figure 4.24, then, is a reasonable guide to vulnerable periods, but clinical data, like those in Figure 4.25, provide an accurate picture for a specific teratogen.

Question 4.5

The eye-closure experiments indicate that, where input from one eye to the cortex is reduced or abolished, the remaining inputs expand to take over neurons previously receiving input from the closed eye. This suggests that the fine structure of the visual cortex is determined by competition for synaptic space by different inputs. The reverse-occlusion experiments point to a limited period when this competition can take place. If the reversal is done too late, the newly opened eye cannot take over any inputs to the cortical cells, so the cortical cells never respond to the newly opened eye.

Question 4.6

There are a number of possible reasons why some fetuses were unaffected by the rubella infection, but the most likely are: those fetuses may have been particularly resilient, much as people are affected by flu virus to different extents; differences in immune responsiveness (the fetal immune system is capable of a limited response to infection from about the fifth month of gestation); the level of infection may have been slight; the infection may have been very brief; some other factor may have been present which ameliorated the rubella attack.

Question 4.7

The reason is that the effect of testosterone on the young female is exerted through oestrogen. Testosterone can be converted to oestrogen within the cells, so in both cases oestrogen is the active agent.

Question 4.8

The evidence from the zebra finch suggests that singing requires a two-stage process. In the example quoted, the injection of testosterone into the adult female zebra finch does not result in singing, which suggests that the female zebra finch brain has not been organized.

Question 4.9

(i), (ii) and (iv) are correct.

The HVc is sexually dimorphic in zebra finches (v), being larger in the male. (vi) is meaningless; there is a sexually dimorphic nucleus of the preoptic area (SDN-POA).

Question 4.10

The statement is an over-simplification. Biological determinants of gender identity combine with social determinants of gender identity; the social determinants do not act independently of the biological determinants. The fact that gender is socially *defined* does not mean that an individual's gender identity is socially *determined*.

Chapter 5

Question 5.1

It would suggest that failure of recall is due to active interference between the earlier items and the items of the second test, rather than passive decay of the latter. (In fact, such results have been obtained and they cast some doubt on the decay theory.)

Question 5.2

The hippocampus appears to play a key role in the process of consolidation of memory in mammals, especially that associated with declarative type memories. However, motor skill learning (for which procedural memories are formed) does not necessarily utilize the hippocampus. Likewise the Montreal patient HM (Section 5.7.3) can learn certain motor skills despite the fact that his declarative memory is severely impaired because of the damage to his hippocampus. Memory is not a unitary process.

Question 5.3

The woman has a normal memory for events in the distant past, and a normal memory for events occurring a minute or so before recall; otherwise conversation would have been impossible. The problem was either that she could not consolidate new memories, or that events recently consolidated were inaccessible (hence the failure to recall the near past).

Question 5.4

Figure 5.1 shows memory for nonsense syllables decaying to almost zero in about 20 s, which would suggest that short-term memory only lasts about 20 s. If this were all the memory for recent events that the woman had, conversation would have been almost impossible. Therefore some consolidation must have occurred, but those memories became inaccessible, or were lost, by the next day.

Question 5.5

Evidence for some form of short-term memory in animals comes from studies of passive avoidance learning in chicks. Subconvulsive electric shock given immediately after training causes amnesia for the task; subconvulsive electric shock given more than ten minutes after training does not cause amnesia for the task.

Question 5.6

The correct answer is (d).

Inhibition of protein synthesis means that new proteins are not made. Potassium permeability (a), neurotransmission (b), frequency of action potentials (c), and short-term memory (e) do not require new protein synthesis, and so would be unaffected by the protein synthesis inhibitor. (f) is nonsense, since there is no stable molecule code for memory.

Question 5.7

(a) Morphological changes are likely to occur when long-term memory is formed. The data presented in Section 5.5 on passive avoidance learning in the chick shows that this is the case. But in those experiments there was an *increase* in dendritic spine density following avoidance learning, not a decrease, although the time-scales of the two learning experiments were different. However, it is possible that memory is associated with synaptic loss (Chapter 3). But it cannot automatically be assumed that changes in the density of dendritic spines are correlated precisely with changes in synaptic number. Synapses are found on structures other than dendritic spines (for example cell bodies), so, unless a true estimate is made of total synapse number, such an assertion may not be correct.

(b) Proper control experiments were not carried out. The only valid comparison would have been between trained and untrained chicks of the same age, not those one week younger. Normal developmental changes may result in a decrease in spine density. Also, side effects (such as stress associated with the acoustic learning experience) could have caused the observed changes in dendritic spine density. Controls to eliminate these confounding variables could have involved blockage of memory by drugs or by subconvulsive electric shock, to disrupt memory formation in chicks subject to the acoustic experience. These should then have been compared with the chicks that had been successfully imprinted on the sound.

Question 5.8

Suppose that short-term memory is due to activation of specific neuronal circuits, and that the longer that activation of these circuits occurs, the more effective will be the consolidation into long-term memory. If neurons of the circuit release a neurotransmitter such as acetylcholine and, due to DFP, relatively large amounts are available, this may prolong activation of the circuits. Whether this occurs with acetylcholine is not clear, but you will recall that another neurotransmitter (glutamate) in the hippocampus does indeed bring about activation of neural circuits in the process called long-term potentiation.

Question 5.9

(e) is correct, because it includes all the possible conclusions, that is, (a), (b), (c) and (d). It would not be correct to consider merely one of the possible conclusions as *the* conclusion.

GLOSSARY

caste   All those individuals within a social group that perform a particular set of tasks. (Section 4.3.1)

cell adhesion molecules   Class of substance that attracts growing axons. (Section 2.4.3)

chemotactic guidance   Contact-mediated interactions between axons, which direct growth. (Section 2.4.5)

chemotropic guidance   Process by which a target releases an attractant, which promotes the growth of an axon through the attractant and towards the target. (Section 2.4.5)

chemotrophic factors   Factors that promote cell survival. (Section 3.2.2)

collateral sprouting   The growth of axons out of existing axons, especially in the presence of a denervated target. (Section 3.6.1)

congenital diseases   A disease, either genetic or non-genetic, whose symptoms are already present at the time of birth. (Section 4.5.2)

differentiation phase   The phase of development in which cells form the characteristic features of neurons. (Section 2.1)

extracellular matrix   The environment between cells, which consists of proteins and other substances that are produced by the cells and also the outer membranes of the cells themselves. (Section 2.4.2)

genetic marker   Protein that a particular genotype produces. Used in conjunction with genetic mosaics so that cell types with different genotypes can be visually distinguished. (Section 4.2.1)

growth cone   The tip of the axon or dendrite that grows by extending fingers of cell material. (Section 2.4.1)

gynandromorph   A genetic mosaic animal which has some cells with a female genotype and some cells with a male genotype. (Section 4.2.1)

Hebb synapse   Synapse in which simultaneous activity in the presynaptic and postsynaptic zones strengthens the synapse. (Section 5.4)

induction   The ability of cells or tissue to influence the differentiation of nearby cells, presumably by a chemical signal. (Section 2.3.3)

lineage   The ancestry of the cell. (Section 2.3.1)

long-term potentiation (LTP)   A persistent increase in the responsiveness of neurons to a single electrical pulse. It may be induced by intense (that is, high frequency) electrical stimulation. (Section 5.7.3)

nerve growth factor (NGF)   A factor that stimulates the growth of axons and neurites and promotes neuronal survival. (Section 3.2.2)

neural cell adhesion molecule (NCAM)   A substance occurring on the surface of axons, which assists in their guidance by binding to the same material on other axons. (Section 2.4.3)

neural tube   The hollow tube that develops into the brain and spinal cord. (Section 2.2)

neural crest   Cells that are found on the dorsal part of the neural tube, which later develop into the peripheral nervous system. (Section 2.3.5)

neurites   Dendrites or axons grown from neurons in tissue culture. (Section 2.4.1)

neuronal plasticity   Property of neurons whereby the connection between a neuron and a target is neither specified nor fixed; the ability of neurons to change their connections. (Section 1.3)

neuronal specificity   Property of neurons whereby particular neurons are always connected to particular targets; the inability of neurons to change their connections. (Section 1.3)

open field   Apparatus used to test the activity of animals; a large, brightly lit circular arena with no hiding places. (Section 4.3.2)

passive avoidance learning   A procedure in which learning is said to have occurred when an animal withholds (that is, does not perform) a particular response. (Section 5.3.2)

physical caste   A caste that differs in its physical appearance from other castes. (Section 4.3.1)

regulation   The ability of embryonic tissue to replace missing parts. (Section 2.3.2)

stem cells   Cells that divide and differentiate to become neurons and glia. (Section 1.1)

synaptogenesis   The formation of a synapse between neurons and their targets. (Section 3.1)

temporal caste   A caste distinguished by the fact that the particular group of tasks performed are only performed by animals of a certain age; an animal may belong to different temporal castes at different ages. (Section 4.3.1)

teratogens   External factors that interfere with development, usually to the detriment of the organism. (Section 4.5.3)

tissue culture   A technique that allows living cells to be grown outside the body. (Section 2.4.1)

trauma   Non-specific, unnatural and vigorous stimulation of the brain, for example from accidental blows to the head or electroconvulsive shock. (Section 5.3.2)

ACKNOWLEDGEMENTS

Grateful acknowledgement is made to the following sources for permission to reproduce material in this book:

Figure 1.1: modified from an illustration by Tom Prentiss in Cowan, W. M. (1979) The development of the brain, *Scientific American,* **241**, No. 3, pp. 107–117; *Figure 1.2:* Roberts, D. F. and Thomson, A. M. (1976) *The Biology of Human Fetal Growth*, Taylor and Francis Ltd; *Figure 2.3a:* Schroeder, T. E. (1970), *Journal of Embryology and Experimental Morphology,* **23**, pp. 427–462, The Company of Biologists Ltd; *Figure 2.3b:* Mohun, T., Tilly, R., Mohun, R. and Slack, J. M. W. (1980) *Cell,* **22**, pp. 9–15, Cell Press, Cambridge Ma.; *Figure 2.3c:* Sussman, M. (1964) *Growth and Development*, Prentice Hall; *Figure 2.4:* Hamburger, V. (1977) *Neuroscience Research Program Bulletin*, **15**, Suppl. III, pp. 1–37, MIT Press Journals; *Figure 2.5:* Hirose, G. and Jacobson, M. (1979) *Developmental Biology*, **71**, pp. 191–202, Academic Press; *Figure 2.9:* reprinted by permission from Angevine, J. B. and Sidman, R. L. (1961) *Nature,* **192**, pp. 766–768. Copyright © 1961 Macmillan Magazines Ltd; *Figure 2.13:* Stuermer, C. A. O. (1990) *Neuroscience Research,* Supplement **13**, pp. S1–S10, Elsevier Science Publishers, Ireland Ltd; *Figure 2.14:* Goldberg, D. J. and Burmeister, D. W. (1989) *Trends in Neurosciences,* **12**, Elsevier Science Publishers; *Figure 2.15:* Swanson, G. and Lewis, J. (1986) *Journal of Embryology and Experimental Morphology,* **95**, pp. 37–52, The Company of Biologists Ltd; *Figure 2.16:* Letourneau, P. C. (1975) *Developmental Biology*, **44**, pp. 92–101, Academic Press; *Figure 2.18:* reprinted by permission from Bentley, D. and Caudy, M. (1983) *Nature,* **304**, pp. 62–65. Copyright © 1983, Macmillan Magazines Ltd; *Figure 2.21:* Mattson, M. P., Don, P. and Kater, S. B. (1988) *Journal of Neuroscience*, **8**, 2087–2100, Society for Neuroscience; *Figure 3.1:* adapted from Purves, D. and Lichtman, J. W. (1985), *Principles of Neural Development*, Sinauer Associates; *Figure 3.2:* Hamburger, V. (1934) *Journal of Experimental Zoology,* **68**, pp. 449–494 and Hamburger, V. (1939) *Journal of Experimental Zoology,* **80**, pp. 347–389. Copyright © 1934 and 1939, reprinted with permission from Wiley–Liss, a Division of John Wiley and Sons Inc; *Figure 3.3:* courtesy of R. Levi-Montalcini and V. Hamburger; *Figure 3.4:* adapted from O'Leary, D. D. M., Fawcett, J. W. and Cowan, W. M. (1986) *Journal of Neuroscience,* **6**, pp. 3692–3705, Society of Neuroscience; *Figure 3.5:* Cowan, W. M. (1979), The development of the brain, *Scientific American,* **241**, No. 3, pp. 107–117; *Figures 3.6 and 3.7:* Frank, E., Smith, C. and Mendelson, B. (1988) in Easter, S., Jr, Barald, K. F. and Carlson, B. M. (eds) *From Message to Mind: Directions in Developmental Neurobiology,* Sinauer Associates, Inc*; Figure 3.9:* Sretavan, P. W. and Shatz, C. J. (1986) *Journal of Neuroscience,* **6**, pp. 234–251, Society of Neuroscience; *Figure 3.12*: Courtesy of Professor Jerry Silver; *Figure 3.13:* reprinted with permission from Berry, M., Rees, L., Hall, S., Yin, P. and Sievers, J. (1988) **Brain Research Bulletin**, **20**, 223–231, © copyright 1988 Pergamon Press plc; *Figure 3.15:* adapted from Merrill, E. G. and Wall, P. D. (1978) in Cotman, C. W. (ed.) *Neuronal Plasticity,* © Raven Press, New York; *Figure 3.16:* reprinted with permission from Merzenich, M. M. *et al.* (1983) *Neuroscience*, **10**, pp. 639–665 © copyright 1983, Pergamon Press plc; *Figure 3.17:* adapted from Lichtman *et al.* (1987), *Journal of Neuroscience,* **7**, pp. 1215–1222, Society for Neuroscience; *Figure 3.18:* reprinted by permission from Purves, D. and Hadley, R. D. (1985) *Nature,* **315**, pp. 404–406; *Figure 4.1:* Hotta, Y. and Benzer, S. (1976), *Proceedings of the National Academy of Sciences of the U.S.A.,* **73**, pp. 4154–4158, National Academy of Sciences of the U.S.A.; *Figures 4.3 and 4.6:* Wilson, E. O. (1975) *Sociobiology,* The Belknap Press of Harvard University; *Figure 4.4:* Oster, F. O. and Wilson, E. O., *Caste and Ecology in the Social Insects.* Copyright © 1978 by Princeton University Press, reprinted by permission of Princeton University Press; *Figure 4.5* courtesy of Professor Thomas Eisner, Cornell University; *Figures 4.12 and 4.13:* Knudsen, E. I. (1988) in Easter, S. S., Jr, Barald, K. F. and Carlson, B. M. (eds), *From Message to Mind: Directions in Developmental Neurology,* Sinauer Associates; *Figure 4.16:* Levine, R. B. (1986) *Trends in Neurosciences,* **9**, pp. 315–319; © 1986 Elsevier Science Publishers, B.V.; *Figure 4.17:* Adkins, E. K. (1975), *Journal of Comparative and Physiological Psychology,* **89**, pp. 61–71, copyright 1975 by the American Psychological Association; reprinted with permission; *Figure 4.18:* adapted from Yen, S. S. C. and Jaffe, R. B. (eds) (1978), *Reproductive Endocrinology,* W. B. Saunders and Co. Inc.; *Figure 4.19:* Gorski, R. A. (1988) in Easter, S. S. Jr, Barald, K. F. and Carlson, B. M. (eds), *From Message to Mind: Directions in Developmental Neurology,* Sinauer Associates; *Figure 4.20:* adapted from an illustration by Patricia J. Wynne from Nottebohm, F. (1989), From birdsong to neurogenesis, *Scientific American,* **260**, No.2, pp. 56–61; *Figure 4.21:* Nordeen, E. J. and Nordeen, K. W. (1990), *Trends in Neurosciences,* **13**, pp. 31–36 © 1990 Elsevier Science Publishers; *Figures 4.22, 4.24 and 4.26:* Clarke-Stewart, A. and Friedman, S. (1987), *Child Development: Infancy Through Adolescence.* Copyright © 1987 John Wiley and Sons Inc.; reprinted by permission of John Wiley and Sons Inc.; *Figure 4.25:* Munro, J. D., Sheppard, S., Smithells, R. W., Holzell, H. and Jones, G. (1987) *The Lancet,* **2**, pp. 201–204, © The Lancet Ltd 1987; *Figures 5.4 and 5.7:* courtesy of Dr Michael Stewart; *Figure 5.10a:* courtesy of Professor David Gadian, The Royal College of Surgeons; *Figure 5.10b*: courtesy of Dr Ralph Myers, MRC Cyclotron Unit, Hammersmith Hospital.

INDEX